Alfatih Ahmed Alamin Osman
Investigation of Uranium Binding Forms in
Environmentally Relevant Waters and Bio-fluids

AF891512

TUDpress

Gedruckt mit Unterstützung des Deutschen Akademischen Austauschdienstes

Alfatih Ahmed Alamin Osman

Investigation of Uranium Binding Forms in Environmentally Relevant Waters and Bio-fluids

TUD*press*
2014

Die vorliegende Arbeit wurde am 03. März 2014 an der Fakultät Mathematik und Naturwissenschaften der Technischen Universität Dresden als Dissertation eingereicht und am 04. Juni 2014 verteidigt.

Gutachter:
Prof. Dr. rer. nat. habil. Eckhard Worch
Prof. Dr. rer. nat. habil. Gert Bernhard

Bibliografische Information der Deutschen Nationalbibliothek
Die Deutsche Nationalbibliothek verzeichnet diese Publikation in der Deutschen Nationalbibliografie; detaillierte bibliografische Daten sind im Internet über http://dnb.d-nb.de abrufbar.

Bibliographic information published by the Deutsche Nationalbibliothek
The Deutsche Nationalbibliothek lists this publication in the Deutsche Nationalbibliografie; detailed bibliographic data are available in the Internet at http://dnb.d-nb.de.

ISBN 978-3-944331-72-0

© 2014 TUDpress
Verlag der Wissenschaften GmbH
Bergstr. 70 | D-01069 Dresden
Tel.: 0351/47 96 97 20 | Fax: 0351/47 96 08 19
http://www.tudpress.de

Alle Rechte vorbehalten. All rights reserved.
Gesetzt vom Autor.
Printed in Germany.

Investigation of Uranium Binding Forms in Environmentally Relevant Waters and Bio-fluids

D I S S E R T A T I O N

zur Erlangung des akademischen Grades

Doctor rerum naturalium
(Dr. rer. nat.)

vorgelegt

der Fakultät Mathematik und Naturwissenschaften
der Technischen Universität Dresden

von

M.Sc. Alfatih Ahmed Alamin Osman

geboren am 22.08.1974 **in** Gedarif, Sudan

Eingereicht am 03.03.2014

Die Dissertation wurde in der Zeit von 04/2010 bis 12/2013
im Institut für Ressourcenökologie,
Helmholtz-Zentrum Dresden-Rossendorf e.V. angefertigt

To my parents, with love

Acknowledgements

First, I would like to express my sincere gratitude to Prof. Dr. G. Bernhard for the scientific supervision, for introducing me to this special area in Radiochemistry, and for his unlimited support throughout the whole PhD period.

I am so grateful for my technical supervisor Dr. G. Geipel for his thoughtful guidance in the laboratory, deep discussion and critical feedback on the manuscripts, and for giving me the precious chance to learn from his expertise on Laser Fluorescence Spectroscopy.

I am gratefully acknowledged the German Academic Exchange Service (DAAD) for financial support of the PhD research work and the German language course.

I would like to thank Prof. Dr. E. Worch for his valuable discussion on drinking water quality and treatment, and for reviewing my manuscript.

I am pleased to thank my colleagues L. Lütke, J. Schott, and R. Steudtner for experimental support and C. Joseph for helping with the thermodynamic calculations by EQ3/6.

My gratitude also goes to the lab staff: U. Schäfer, A. Ritter, and C. Eckardt for ICP-MS, IC, and TOC measurements. Special thanks and appreciation to G. Grambole for the invaluable laboratory assistance.

Thanks are due to Prof. F. Winde for providing groundwater samples and to the volunteers at HZDR for providing the body fluids specimens.

My heartfelt thanks and appreciation are due to my wife, my kids, all friends and families in Sudan for their continuous support and encouragement.

Alfatih A. A. Osman
Dresden, Germany
February 2014

Contents

List of abbreviations and symbols

Abbreviations:

Amy	Alpha amylase
Cit	Citrate
EC	Electrical conductivity
Eh^{o}	Standard redox potential
FWHM	Full Width at Half maxima
Gly	Glycine
IC	Ion chromatography
ICCD	Intensified Charge-Coupled Device
ICP-MS	Inductively Coupled Plasma – Mass Spectrometry
Lac	Lactate
MEDUSA	Make Equilibrium Diagrams Using Sophisticated Algorithms
Nd:YAG	Neodymium-Doped Yttrium Aluminum Garnet
NEA-TDB	Nuclear Energy Agency – Thermochemical Database
Ox	Oxalate
SPECFIT	Spectral Fitting Tool for the calculation of stability constants from spectroscopic data
TIC	Total Inorganic Carbon
TOC	Total Organic Carbon
TRLFS	Time-Resolved Laser-Induced Fluorescence Spectroscopy
U_{nat}	Natural uranium

Symbols:

K	Kelvin
Log β	Complex Stability constant
λ_{ex}	Excitation Wavelength
λ_{em}	Emission Wavelength
p	Partial pressure
τ	Fluorescence lifetime

List of publications

Osman AAA, Geipel G, Bernhard G, Worch E (2013) Investigation of uranium binding forms in selected German mineral waters. *Environmental Science and Pollution Research* 20:8629-8635

Osman AAA, Geipel G, Bernhard G (2013) Interaction of uranium(VI) with bioligands present in human biological fluids: the case study of urea and uric acid. *Radiochimica Acta* 101:139-148

Wang Y, Frutschi M, Suvorova E, Phrommavanh V, Descostes M, **Osman AAA**, Geipel G, Bernier-Latmani R (2013) U(IV) release from a mining-impacted wetland. *Nature Communications* 4:2942 doi: 10. 1038/ncomms3942

Osman AAA, Geipel G, Barkleit A, Bernhard G (*in preparation*) Uranium(VI) binding forms in some human body fluids: spectroscopy and modeling investigation

Summary

Uranium from natural and anthropogenic sources is continuously and increasingly released into the environment, enters the food chain, and eventually accumulates into the bio-systems including humans. Therefore, uranium exposure to human becomes more and more a subject of interest in many scientific disciplines such as environmental medicine, toxicology, and radiation protection. Ingestion of uranium via food and drinking water is one of the important pathways of exposure. The ongoing discussion in Germany and the European Union (EU) on the maximum uranium levels in drinking water reflects the increasing awareness of uranium as a public health problem. For water suppliers in many countries around the world, uranium contamination of drinking water also becomes a salient issue.

Nevertheless, the health consequences are dependent not only on the level and duration of exposure, but mainly on the physical and chemical forms of uranium (speciation). Unlike total concentration, speciation can provide more information on uranium behavior (mobility, bioavailability, and toxicity) in both geo-systems and bio-systems. While uranium speciation in natural aquatic systems has been covered quite well in the field of radiochemistry, knowledge is still limited about natural waters with low uranium concentrations. Furthermore, there is a demand for experimental data on uranium speciation in human body fluids which can be of great importance not only to understand its biokinetics, but also relevant for risk assessment and for designing decorporation therapy in the case of accidental overexposure.

The aim of the current work is to elucidate uranium speciation in natural samples (water, bio-fluids) by both spectroscopic means and theoretical modeling. For this purpose, environmentally relevant waters (mineral water, groundwater, acid drainage waters) and natural human body fluids have been investigated. The cryogenic time-resolved laser-induced fluorescence spectroscopy (cryo-TRLFS) at 153 K has successfully been used to identify the speciation in such complex matrices of high organic content and low uranium concentrations without any prior treatment. Speciation calculations using the most recent thermodynamic stability constants for aqueous uranium complexes were also used in line with the spectroscopy.

Among the investigated waters, much focus was given to the mineral waters as a challenge to get detailed information about uranium speciation at *extremely low concentrations* (< 2.0 μg/L). For this purpose, selected mineral waters (CLASSIC type: pH 5 – 6, and STILL type: pH 6.5 – 7.5) from German producers were investigated. The spectroscopic data showed a prevalence of the aquatic species $Ca_2UO_2(CO_3)_3$ in all investigated mineral waters, while other uranyl–carbonate complexes, *viz*, $UO_2CO_{3(aq)}$ and $UO_2(CO_3)_2^{2-}{}_{(aq)}$ existed as minor species in CLASSIC waters only. These complexes were demonstrated from the bi-exponential decay function which was used to fit the luminescence decays of uranium species in the CLASSIC waters. The analysis of the individual emission peaks which showed a spectral shift with increasing delay time also confirmed a second species. The minor species was demonstrated by comparison with spectral data of uranyl carbonate model solutions and was assigned as $UO_2CO_{3(aq)}$ complex. The pH value, alkalinity (CO_3^{2-}), and the main water inorganic constituents, specifically the Ca^{2+} concentration, showed a clear influence on uranium speciation.

The groundwater samples with *exceptionally high U content* from Pofadder, South Africa showed luminescence decays of uranium species that were best fitted with a mono-exponential decay function indicating the sole presence of the ternary species $Ca_2UO_2(CO_3)_{3(aq)}$ as in mineral waters of the STILL type. Thus, the elevation in uranium concentration did not influence its binding forms in the pH range of drinking waters. The modeling results for the main uranium binding forms in both mineral waters and groundwater indicated a good agreement with the spectroscopic measurements. In the case of acid drainage waters, thermodynamic prediction demonstrated the dominance of uranium sulfate complexes which was not consistent with the cryo-TRLFS data. The spectral information at 153 K for both acid drainage water samples indicated no U(VI) – sulfate species, but mainly the luminescence characteristics of U(VI) – chloride system, which possesses a characteristic spectral feature with high luminescence yield even much greater than that of U(VI) complexes with SO_4^{2-}, PO_4^{3-}, and OH^-.

The time-resolved laser-induced fluorescence spectroscopy at room temperature (298 K) was also applied for the complexation of uranium(VI) with bioligands that present in human biological fluids, *viz*, urea and uric acid in aqueous solution at pH (3 for uric acid; 4 for urea) and *I*: 0.1 M ($NaClO_4$). In both complex systems a static quench effect with increasing ligand concentration and no peaks shift upon complexation were observed. With uranium(VI) both ligands formed a fairly weak 1:1 complex with average stability constants of log β_{110} = 4.67±0.29 for uric acid and log β_{110} = 3.79±0.15 and 2.12±0.18 for relatively low and relatively high urea concentrations, respectively. The newly generated data was successfully applied on the U(VI) speciation modeling in body fluids, *e.g.*, human urine and sweat.

For the first time, experimental data on U(VI) speciation in *real natural body fluids* (saliva, sweat, urine) was obtained by means of cryo-TRLFS at 153 K. By using the time dependency of fluorescence decay and the band positions of the emission spectra, various uranyl complexes were demonstrated in the studied samples that were previously spiked with U(VI). The variations of the body fluids in terms of chemical composition, pH, and ionic strength resulted in different binding forms of U(VI). The bioorganic ligands influenced the speciation of U(VI) in both saliva and urine, whereas in sweat the speciation was governed mainly by inorganic ligands.

The speciation of U(VI) in human saliva was mostly affected by the presence of the enzyme α-amylase. The spectral maxima of the pure α-amylase-U(VI) system over the pH range 4 to 7 were somewhat similar to that of natural saliva. The slight difference was attributed to the complexity of the natural saliva and to the variation of α-amylase secretion and concentrations from that of the pure system. The degree of similarity is rather consistent with the theoretical speciation when α-amylase was included in the equation model, where the complex $UO_2[(COO)_2\text{-}Amy]^{2+}$ predominated until around pH 6.6.

At the pH of the studied human sweat, the luminescence characteristics of U(VI) indicated the aqueous $Ca_2UO_2(CO_3)_3$ which is the typical binding form in natural waters. Thus, uranium in human sweat behaves like in natural waters and was not affected by bioligands (urea, uric acid, lactic acid) that are secreted by sweat glands. The thermodynamic modeling approved the same U(VI) species as indicated by fluorescence measurement.

In contrast, U(VI) in natural human urine under investigation (pH 5 – 6) showed different luminescence characteristics from that in saliva and sweat. This is mainly attributed to the matrix complexity of natural human urine (e.g. organics, ionic strength). The emission band maxima of natural urine are comparable but not exactly similar to that of pure systems of UO_2^{2+}/Cit^{3-}. Speciation modeling supported the spectroscopic data and demonstrated $UO_2(Cit)^-$ as the main U(VI) binding form in natural urine samples (pH range 5 – 6).

With regard to inorganic ligands, phosphates and carbonates as biological buffers ($H_2PO_4^-$ /HPO_4^{2-}, HCO_3^-/CO_3^{2-}) showed a great influence on U(VI) speciation in the body fluids especially in the presence of Ca^{2+} ions. For instance, in carbonate-poor urine the prevailing complex $Ca(UO_2)_2(PO_4)_{2(aq)}$ which is sparingly soluble above pH 7.5 and can form a fine precipitate with the time could give a new perspective for bioavailability of uranium within the renal system. By using thermodynamic data for aqueous uranium complexes that are compiled in the NEA database and in some recent literature, chemical thermodynamic calculations of the main U(VI) species in the investigated body fluids provided a good estimation for U(VI) speciation which is somehow consistent with spectroscopic data.

The results of the present work add new data regarding the uranium(VI) speciation in both natural waters and bio-fluids. Data about uranium binding forms in natural waters could be used for more insight research on uranium chemo-toxicity in drinking water since there is no available information on the long-term exposure of *low uranium concentrations* via ingestion on humans. Altogether the obtained data provides a comprehensive knowledge on uranium(VI) speciation in natural waters and bio-fluids which offers a reliable basis and a better understanding as a basic mechanism of redistribution of uranium after exposure, and thus for more insight research on its biokinetics. Furthermore, the suitability of the cryo-TRLFS as a non-invasive technique for direct determination of uranium speciation in such complex matrices with low uranium content was demonstrated.

Zusammenfassung

Aus natürlichen und anthropogenen Quellen wird Uran kontinuierlich und zunehmend verstärkt in die Umwelt freigesetzt, gelangt in die Nahrungskette und akkumuliert sich letztendlich in der Biosphäre, einschließlich im menschlichen Körper. Die mögliche Beeinflussung der menschlichen Gesundheit durch Uran ist zunehmend Gegenstand der Forschung in vielen Wissenschaftsgebieten, wie Umweltmedizin, Toxikologie und Strahlenschutz, geworden. Die Aufnahme von Uran über die Nahrung und Trinkwasser ist dabei einer der wichtigsten Pfade zur Inkorporation von Uran. Die fortlaufende Diskussion in Deutschland und der Europäischen Union (EU) zu einen Maximalwert von Uran im Trinkwasser, spiegelt diese Aktualität wieder. Für die Wasserversorger in vielen Ländern der Erde, ist die Urankontamination von Trinkwasser ein herausragendes Problem geworden. Jedoch sind die gesundheitlichen Auswirkungen nicht nur abhängig von der Höhe und Dauer der Uranaufnahme, sondern hauptsächlich auch von den physikalischen und chemischen Formen des Urans (Speziation). Anders als die Angabe der Totalkonzentration, liefert die Speziation mehr Informationen zum Verhalten von Uran in Bio- und Geosystemen (Mobilität, Bioverfügbarkeit und Toxizität).

Obwohl die Uranspeziation in natürlichen wässrigen Systemen schon recht gut auf dem Gebiet der Radiochemie behandelt wurde, ist das Wissen über natürliche Wässer mit niedrigen Urankonzentrationen noch immer begrenzt. Weiterhin besteht der Bedarf an experimentellen Daten zur Uranspeziation in menschlichen Körperflüssigkeiten, welche von größter Wichtigkeit sein können, nicht nur um Biokinetiken zu verstehen, sondern auch um im Fall eines nuklearen Unfalls und Inkorporation von Uran in den menschlichen Organismus eine Risikobewertung vornehmen zu können und Therapien für die Entgiftung zu entwickeln.

Das Ziel dieser Arbeit ist die Aufklärung der Uranspeziation in natürlichen Proben (Wasser, Biofluide) durch spektroskopische Methoden und theoretische Modellierung. Zu diesem Zweck wurden umweltrelevante Wässer (Mineralwasser, Grundwasser, saures Drainagewasser) und natürliche menschliche Körperflüssigkeiten untersucht. Die zeitaufgelöste laserinduzierte Fluoreszenzspektroskopie bei Tieftemperatur (cryo-TRLFS) wurde erfolgreich eingesetzt um die Speziation in komplexen Matrizes mit hohem Organikanteil und niedrigen Urankonzentrationen ohne Vorbehandlung der Probe aufzuklären. Speziationen unter Nutzung der aktuellsten thermodynamischen

Stabilitätskonstanten für wässrige Urankomplexe wurden ebenfalls berechnet und mit den spektroskopischen Ergebnissen verglichen.

Ein besonderer Schwerpunkt wurde auf die Untersuchung von Mineralwässern gelegt um genaue Informationen zur Uranspeziation bei *sehr niedrigen Konzentrationen* (< 2,0 µg/L) zu erlangen. Dazu wurden ausgewählte Mineralwässer (CLASSIC-Typ: pH 5-6; STILL-Typ: pH 6,5-7,5) von deutschen Herstellern untersucht. Die spektroskopischen Daten zeigten die stark vorherrschende aquatischen Spezies $Ca_2UO_2(CO_3)_{3(aq)}$ in allen untersuchten Mineralwässern, während andere Uranylcarbonatkomplexe, wie $UO_2CO_{3(aq)}$ und $UO_2(CO_3)_2^{2-}$, als untergeordnete Spezies ausschließlich in Mineralwässern vom CLASSIC-Typ vorkamen. Diese Komplexe wurden aus einer biexponentiellen Zerfallsfunktion, welche für den Fit des Lumineszenzzerfalls der Uranspezies in den CLASSIC Wässern verwendet wurde, nachgewiesen. Die Analyse der einzelnen Emissionsmaxima, welche eine spektrale Verschiebung mit steigender delay time zeigten, bestätigt ebenfalls eine zweite Spezies. Die untergeordnete Spezies wurde durch den Vergleich mit Spektren von Uranylcarbonat-Modelllösungen ermittelt und dem $UO_2CO_{3(aq)}$ Komplex zugeordnet. Der pH-Wert, Alkalinität (CO_3^{2-}) und die anorganischen Hauptwasserbestandteile, insbesondere die Ca^{2+}-Konzentration, zeigten einen deutlichen Einfluss auf die Uranspeziation.

Die Grundwasserproben mit *außerordentlich hohem U-Gehalt* aus Pofadder, Südafrika, wiesen Lumineszenzzerfälle der enthaltenen Uransspezies auf, die am besten mit einer monoexponentiellen Zerfallsfunktion angefittet werden konnten. Diese zeigten die alleinige Präsenz der ternären Spezies $Ca_2UO_2(CO_3)_{3(aq)}$ in den Grundwasserproben, wie auch in den Mineralwässern des STILL-Typs. Demnach beeinflusst ein Anstieg der Urankonzentration nicht die Bindungsform des Urans im pH-Bereich von Trinkwässern. Die Modellierung der Hauptbindungsform des Urans in beiden Mineralwässern und dem Grundwasser zeigte eine gute Übereinstimmung mit den spektroskopischen Messungen. Im Fall der sauren Drainagewässer zeigte die thermodynamische Vorhersage das Vorherrschen von Uransulfatkomplexen an. Diese stimmte jedoch nicht mit den Daten der cryo-TRLFS überein. Die spektroskopischen Informationen bei 153 K für die sauren Drainagewasserproben zeigten keine U(VI)-Sulfatspezies, sondern wiesen vorallem die Lumineszenzcharakteristik des Uranylchloridsystems auf, welches eine charakteristisch hohe Lumineszenzausbeute (viel höher als von U(VI)-Komplexen mit SO_4^{2-}, PO_4^{3-} und OH^-) besitzt.

Die zeitaufgelöste laserinduzierte Fluoreszenzspektroskopie bei Raumtemperatur (298 K) wurde zur Bestimmung der Komplexierung von Uran(VI) mit Bioliganden (Harnstoff und Harnsäure), welche in menschlichen biologischen Flüssigkeiten vorkommen, eingesetzt. Die Komplexierungsuntersuchungen wurden in wässriger Lösung bei I = 0,1 M ($NaClO_4$) und pH 3 (Harnsäure-System) bzw. pH 4 (Harnstoff-System) durchgeführt. In beiden Komplexierungssystemen wurde ein statischer Quencheffekt mit steigender Ligandkonzentration, aber keine Verschiebung in den Bandenlagen, durch Komplexierung beobachtet. Mit Uran(VI) bilden beide Liganden einen schwachen 1:1-Komplex mit durchschnittlichen Stabilitätskonstanten von log β_{110} = 4,67 ± 0,29 für Harnsäure und log β_{110} = 3,79 ± 0,15 für relativ niedrige bzw. log β_{110} = 2,12 ± 0,18 für relativ hohe Harnstoffkonzentrationen. Die Anwendung dieser neu generierten Daten auf die U(VI)-Speziationsmodellierung in Körperflüssigkeiten, z.B. in menschlichen Urin und Schweiß, wurde erfolgreich angewandt.

Erstmals wurden mit der cryo-TRLFS bei 153 K experimentelle Daten zur U(VI)-Speziation in echten natürlichen Körperflüssigkeiten (Speichel, Schweiß, Urin) erhalten. Mit den zeitaufgelösten und/oder spektralen Informationen der Emissionspektren konnten, nach Uranzugabe zu den untersuchten Proben, verschiedene Uranylkomplexe nachgewiesen werden. Durch die chemische Zusammensetzung, pH und Ionenstärke in den jeweiligen Körperflüssigkeiten wird die Bindungsform von U(VI) beeinflusst. Die bioorganischen Liganden beeinflussen vorallem die Speziation von U(VI) im Speichel und Urin, wohingegen im Schweiß die U(VI)-Speziation hauptsächlich von anorganischen Liganden bestimmt wird.

Die Speziation von U(VI) im menschlichen Speichel wurde besonders durch das Enzym α-Amylase beeinflusst. Die Emissionsmaxima in den Spektren des α-Amylase-U(VI) Systems im pH Bereich 4-7 waren einigermaßen ähnlich zu denen in den Spektren vom natürlichen Speichel. Die gewissen Unterschiede zum natürlichen System werden durch die Matrixkomplexität des natürlichen Speichels und durch Variationen in der α-Amylase-Sekretion und –Konzentration verursacht. Die Vergleichbarkeit der gemessenen mit der theoretischen Speziation ist gegeben. Der Komplex $UO_2[(COO)_2\text{-Amy}]^{2+}$ ist die bestimmende U(VI)-Spezies bis etwa pH 6,6.

Bei dem pH des untersuchten menschlichen Schweißes, zeigte die Lumineszenzcharakteristik von U(VI) die wässrige Spezies $Ca_2UO_2(CO_3)_{3(aq)}$ an, welche die typische U(VI)-Bindungsform in natürlichen Wässern ist. Demnach verhält sich Uran im menschlichen Schweiß wie in natürlichen Wässern und dessen Speziation wird nicht durch Bioliganden (Harnstoff, Harnsäure, Milchsäure), die über Schweißdrüsen ausgeschieden werden, beeinflusst. Die Modellierung bestätigt dieselbe U(VI)-Spezies, die fluoreszenzspektroskopisch im Schweiß gefunden wurde.

Im Gegensatz dazu zeigte U(VI) in natürlichen menschlichen Urinproben (pH 5-6) eine von Speichel und Schweiß verschiedene Lumineszenzcharakteristik. Dies wird hauptsächlich der Matrixkomplexität vom natürlichen menschlichen Urin (Organik, Ionenstärke) zugeschrieben. Die Maxima der Emissionsbanden in den Lumineszenzspektren vom natürlichen Urin sind vergleichbar, jedoch nicht exakt gleich, mit denen in den Lumineszenzspektren des reinen UO_2^{2+}/Cit^{3-} Systems. Die Speziationsmodellierung unterstützt die spektroskopischen Daten und gibt die Spezies $UO_2(Cit)^-$ als Hauptbindungsform von U(VI) in den natürlichen Urinproben (pH 5-6) an.

Bezüglich der anorganischen Liganden haben Phosphate und Carbonate, die als biologische Puffer ($H_2PO_4^-/HPO_4^{2-}$, HCO_3^-/CO_3^{2-}) wirken, vorallem in Anwesenheit von Ca^{2+} einen großen Einfluss auf die Speziation von U(VI) in Körperflüssigkeiten. So überwiegt beispielsweise der Komplex $Ca(UO_2)_2(PO_4)_{2(aq)}$ in carbonatarmen Urin. Dieser Komplex ist über pH 7,5 schwerlöslich und bildet mit der Zeit in Lösung ein feines Präzipitat aus. Damit muss die Bioverfügbarkeit von Uran innerhalb des renalen Systems neu bewertet werden. Die Modellierung der Hauptspezies von U(VI) in den untersuchten Körperflüssigkeiten brachte eine Speziation von U(VI) hervor, die einigermaßen mit den spektroskopischen Daten übereinstimmt.

Die Ergebnisse dieser Arbeit liefern neue Daten bezüglich der U(VI)-Speziation in natürlichen Wässern und Biofluiden. Die Daten zu den Bindungsformen des Urans in natürlichen Wässern können für weitere Forschungsarbeiten bezüglich der Chemotoxizität von Uran auf den menschlichen Organismus genutzt werden. Bisher gibt es keine verfügbaren Informationen zu den Auswirkungen einer langzeitigen Exposition mit niedrigen Urankonzentrationen durch die Nahrungsaufnahme. Insgesamt liefern die erhaltenen Daten einen umfassenden Wissensstand zur Uran(VI)-Speziation in natürlichen Wässern und Biofluiden, welcher eine verlässliche Basis und ein besseres Verständnis der grundlegenden Mechanismen zur Umverteilung und Biokinetik von Uran nach dessen Aufnahme bietet. Des Weiteren konnte die Eignung der cryo-TRLFS als nicht-invasive

Methode zur direkten Bestimmung der Uranspeziation in komplexen Martices mit niedrigen Urangehalten verdeutlicht werden.

1. Motivations and Objectives

Uranium ubiquitously gets into our environment through different sources and is found in different levels in both abiotic and biotic systems. The dominant reasons for increasing emission of uranium in the environment arise from the use of uranium as a fuel for the generation of nuclear energy and as a basic material for nuclear weapons production, which also includes the use of depleted uranium metal in armor-piercing ammunition. So far, approximately two million tons of uranium have been produced worldwide and currently this total amount increases by around 40.000 tons per year. Nevertheless, it is important to account for the amount of uranium that is emitted into the bio-sphere and absorbed or ingested by organisms as a result of the huge production rates of industrial products such as phosphate fertilizers, cement, potable and mineral waters, and the use of coal. The uranium content of soils, rocks, water, air, and bio-systems also shows a strong dependency on the local environment. Properties such as mobility, bioavailability or toxicity of a chemical element depend not only upon total concentration, but also on the chemical binding form or the so-called speciation. According to that, cell toxicity of uranium was experimentally tested in rats' kidneys and was put in the following order; $UO_2PO_4^- \sim UO_2HPO_4 >> UO_2Cit_2^{4-} > UO_2(CO_3)_3^{4-} > Ca_2UO_2(CO_3)_3$ (Carriere et al. 2004). Changes in physicochemical conditions strongly influence the formation of different uranium complexes/species in aquatic systems or on surface layers, where the important controlling factors of speciation are the pH value, redox potential, ionic strength, and availability of inorganic and organic ligands, solubility product, and the formation of colloids.

However, no detailed information is available about the chemical binding forms (speciation) of uranium in mineral and potable waters, food stuffs, bio-fluids, and the relevant bio-compartments at relevant concentrations. With the present increase of uranium accumulation in the food chain and eventually in the bio-system including humans, knowledge about uranium binding forms with bioligands present in biological fluids at natural relevant concentrations is needed and will be useful to understand its toxicity, behaviour, and distribution in such cellular levels, as well as for evaluating toxicity after uranium exposure. There is a lack of pertinent data to support thermodynamic calculations of U(VI) in human biological fluids, especially with such relevant biological ligands. It would be more meaningful when considering these organic substances even with their weak complexes in the equation model for a reliable and exact prediction. Therefore, considering the interaction of uranium(VI) with the relevant biological ligands is of great importance for an accurate model and/or a better understanding of uranium biokinetics. Previous research about uranium(VI) binding forms in the body fluids was limited to thermodynamic speciation calculations (Sutton and Burastero, 2004), and there is no spectroscopic verification of the obtained data. Moreover, samples were also limited to simulated solutions which contained significantly less constituents than natural aqueous system or body fluids. It could be possible to learn more from nature by the use of spectroscopic identification of unknown species in real samples obtained from a human body.

The aims of this study are:

- To determine uranium binding forms in different natural waters with special focus on mineral waters with *extremely low concentrations* of uranium (< 2.0 µg/L), and to compare the findings with the uranium thermodynamic modeling results of these waters.
- Spectroscopic characterizations of uranium(VI) complexes with some bioligands commonly exist in human body fluids, namely, urea and uric acid. With emphasis to include their associate complex formation constants in the speciation modeling of U(VI) in body fluids (e.g. urine).
- To compare the calculated thermodynamic data of uranium binding forms based on the newest data compilations with the data obtained from spectroscopic measurements of the simulated and original body fluids (saliva, sweat, urine). The challenge is to provide experimental data about U(VI) speciation in such complicated matrices in line with theoretical thermodynamic calculations.

2. Background

2.1 Properties and occurrence of uranium

Uranium is a heavy, silvery-white, ductile, highly-electropositive, and slightly-paramagnetic metal with a poor electrical conductivity (Závodská et al. 2008). It was first discovered in 1789 by the analytical chemist Martin Klaproth during his investigation on a pitchblende sample from Johanngeorgenstadt, near Freiberg, Germany (Meinrath 1998). Uranium is a member of the actinide series in the periodic table, with atomic number 92 and electron configuration; [Rn] $7s^2$ $6d^1$ $5f^3$. The +6 oxidation state (U(VI)) has configuration [Rn] $7s^0$ $6d^0$ $5f^0$ (Cordfunke 1969). As it is very susceptible to oxidation by air and becomes coated with a layer of oxide, uranium occurs in nature mainly in oxidized forms. With its high density (19.05 g/cm^3), uranium is considered as one of the heaviest elements that naturally occur in the earth's crust with a mean concentration of 3 mg/Kg (0.0003%) (ATSDR 1999). This abundance can roughly be compared to that of Mo and As, but at the same time greater than that of Cu, Sb, W, and Cd (Závodská et al. 2008).

Similar to many other heavy metals, uranium forms a wide diversity of minerals including oxides (e.g., pitchblende and uraninite), phosphates, carbonates, vanadates, silicates, arsenates, and molybdates (Clark et al. 1997); most of them are fluorescent when excited with UV radiation (Geipel 2006). Phosphate deposits also contain considerable amount of uranium ranges 50 – 300 mg/Kg and in top soils varies from about 0.1 – 20 mg/Kg with a mean of 2.8 mg/Kg worldwide (UNSCEAR 2000a). Besides that, it occurs at various concentrations in all environmental compartments; abiotic and/or biotic (e.g., air, soil, water, sediments, microorganisms, plants, etc.). Nevertheless, uranium content in the natural environment has been increasingly enhanced by the huge uranium production and processing (mining, milling, and purification) as a nuclear fuel for energy generation or for nuclear weapons manufacturing. Other non-nuclear industries, such as fertilizer processing and use, the mineral water industry, cement industry, oil and coal mining, etc., can also cause high uranium input in the environment. However, it can also be released naturally during weathering of soils and rocks and due to its accumulation in redox fronts (Lichtner and Waber, 1992; Yoshida et al. 2008).

In nature uranium is composed of a mixture of three isotopes, all of which are radioactive: ^{238}U (99.284%), ^{235}U (0.711%), and ^{234}U (0.005%) by mass, and ^{238}U (49%), ^{235}U (2%), and ^{234}U (49%) by radioactivity (ATSDR 1999). Both ^{238}U and ^{235}U are parent radionuclides for two of the three independent natural radioactive decay series, while ^{234}U is a daughter nuclide of the ^{238}U series. All the three isotopes are alpha emitters with long half-lives and natural activity of 25.40 Bq/mg (decays per s and mg U_{nat}), and it is ranked as a weakly radioactive element.

$$^{238}_{92}U = {}^{234}_{90}Th + {}^{4}_{2}He^{2+} \text{ (alpha particle)} \qquad (2-1)$$

Table 2-1. Radiological properties of natural uranium (U_{nat}) isotopes

Isotope	Physical half-life (years)	Relative abundance (%)	Activity (Bq/mg U_{nat})
^{234}U	2.450 x 10^5	0.005	12.4
^{235}U	7.037 x 10^8	0.711	0.6
^{238}U	4.468 x 10^9	99.284	12.4
Total			25.4

2.2 Aqueous chemistry of uranium

In aqueous solution, uranium exists in two major oxidation states; U(IV) and U(VI), that are environmentally relevant. Other valence states are of only minor importance, for instance uranium(V) in aqueous system disproportionates rapidly to a mixture of uranium(IV) and uranium(VI) as presented in Eq.(2-2):

$$2UO_2^{+}{}_{(aq)} + 4H^{+}{}_{(aq)} = UO_2^{2+}{}_{(aq)} + U^{4+}{}_{(aq)} + 2H_2O_{(l)}\,; \quad E_h^{0} = +0.56 \text{ V} \qquad (2-2)$$

Another example is uranium(III) which plays no role in environmental chemistry of uranium since it is only stable in a non-aqueous, strong reducing medium (Geipel 2005a), its compounds in water tend to readily oxidized to uranium(IV) as shown in Eq.(2-3) (Grenthe et al. 1992).

$$U^{4+}{}_{(aq)} + e^{-} = U^{3+}{}_{(aq)}\,; \quad E_h^{0} = -0.553 \text{ V} \qquad (2-3)$$

Therefore, under environmental conditions of natural aquatic systems only tetravalent and hexavalent states are stable. Under those conditions U(IV) is extremely insoluble, thus the solution behavior is dominated by U(VI). Equation (2-4) shows the redox reaction between the two oxidation states (Grenthe et al. 1992).

$$UO_2^{2+}{}_{(aq)} + 4H^+ + 2e^- = U^{4+}{}_{(aq)} + 2H_2O_{(l)}; \quad E_h^{\,0} = +0.268V \qquad (2-4)$$

This great notable difference in solubilities between the two oxidation states (VI) and (IV) determines the behavior of uranium weather it is sorbed, mobilized, precipitated or immobilized in a given aqueous system (Meinrath 1998). Changes in environmental conditions (e.g. pH, *Eh*, salinity, etc.) in natural systems not only influence the oxidation states of uranium, but also the chemical binding form or the so called speciation (see Fig.2-1).

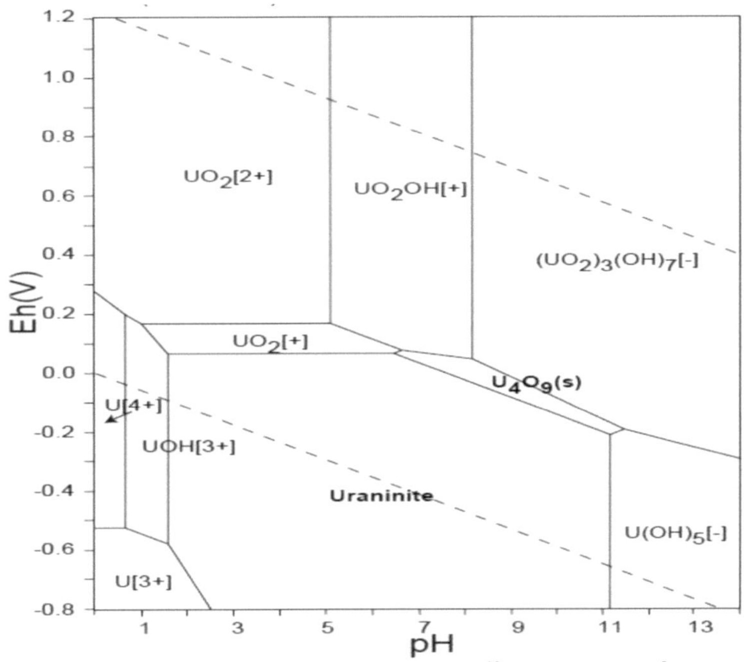

Figure 2-1. Eh-pH diagrams of the system U-O-H. [U]: 10^{-10} M, 298.15 K, 10^5 Pa (Takeno 2005)

2.2.1 Chemical speciation in solutions

In order to understand the complicated chemistry of uranium in natural aquatic environment, we need to have detailed information about uranium speciation in aqueous solution. The term "speciation" was introduced from biology to explain the element's actual molecular form in order to understand its environmental behavior and toxicity. But generally, speciation of an element is the distribution of that element amongst defined chemical species in a system (Templeton et al. 2000). The chemical speciation identifies not only the molecular formula, charge, and structure of the element's compound, but also the isotopic composition of the molecule and the oxidation states of its constituents (Runde 2000). As uranium is one of the most environmental contaminants, its speciation is of great importance giving a comprehensive knowledge for predicting its migration behavior in both geo-systems and bio-systems. In addition, it provides valuable information for remediation strategies on contaminated sites (old uranium mines) and/or designing concept for waste repositories. Furthermore, it is critical to understand toxicity, bioavailability, and environmental fate and transport. For instance the aqueous species $Ca_2UO_2(CO_3)_3$ is a non-bioavailable chemical form for human body and is considered less-toxic comparing to $UO_2(CO_3)_3^{4-}$, while $UO_2Cit_2^{4-}$ is known as a more cyto-toxic chemical form (Carriere et al. 2004). Uranium speciation in natural system is controlled by some environmental factors that are summarized as follows:

- Element concentration
- Temperature
- Pressure (gas equilibria)
- pH value (H^+ concentration)
- Ionic strength (activity coefficient)
- Redox potential (O_2 content)
- Organic and inorganic ligands
- Colloids
- Solid phases
- Type and structure of the available surfaces
- Microbial activities (e.g. bacteria, algae, fungi)

Thermodynamic laws relate these factors to each other and govern uranium speciation in nature.

2.2.2 U(VI) species with inorganic ligands

In oxidizing aqueous systems, uranium(VI) presents as the linear dioxo cation or the so-called uranyl ion $(O=U=O)^{2+}$ which dominates at pH ≤ 2. Depending on pH and uranium concentration, uranyl (UO_2^{2+}) entity undergoes a number of hydrolysis reactions. In the pH range above 3.5 till pH 14, uranium(VI) forms several mono-nuclear and poly-nuclear species: UO_2OH^+, $UO_2(OH)_{2(aq)}$, $(UO_2)_2(OH)_2^{2+}$, $(UO_2)_3(OH)_5^+$, $(UO_2)_4(OH)_7^+$, $(UO_2)_3(OH)_7^-$, $UO_2(OH)_3^-$, $UO_2(OH)_4^{2-}$ (Guillaumont et al. 2003) as illustrated in figure (2-2(a)).

Under atmospheric conditions, the U(VI)/hydrolyses systems will be in equilibrium with atmospheric CO_2 ($p(CO_2)$: $10^{-3.5}$ bar) and eventually various $U(VI)/H_2O/CO_3^{2-}$ will be formed (see Fig.2-2(b)). With raising carbonate concentrations, mono-, di-, and tri-nuclear uranyl carbonate species: $UO_2CO_{3(aq)}$, $UO_2(CO_3)_2^{2-}$, $UO_2(CO_3)_3^{4-}$ become increasingly more important than hydrolytic species. However, the uranyl/CO_3^{2-} systems change in the presence of divalent cations (Me^{2+}) such as Ca^{2+}, Mg^{2+}, Sr^{2+}, and Ba^{2+} forming highly mobile complexes with the formulae $MeUO_2(CO_3)_3^{2-}$ and $Me_2UO_2(CO_3)_{3(aq)}$ (Dong and Brooks, 2006; Geipel et al. 2008).

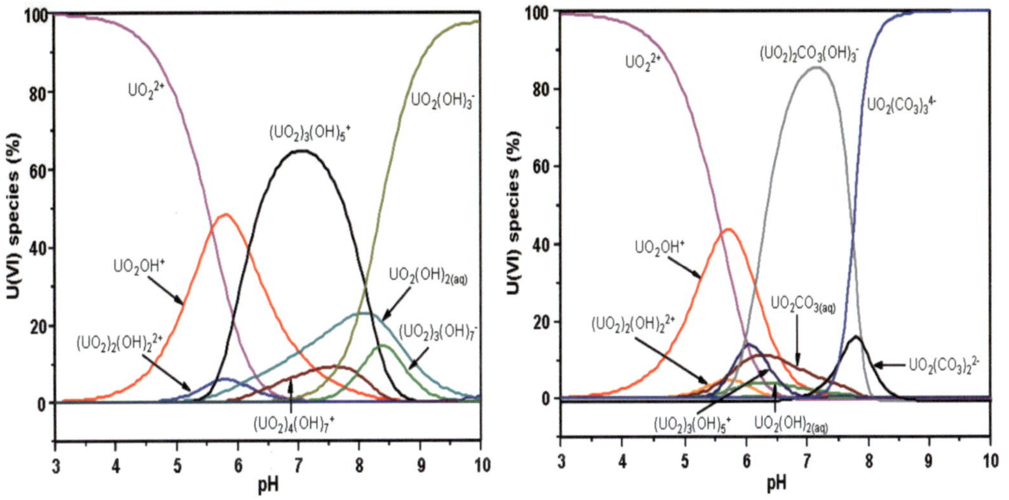

(a) (b)

Figure 2-2. U(VI) binding forms in aquatic solutions in the **(a)** absence of CO_2 and **(b)** $p(CO_2)$: $10^{-3.5}$ bar, and [U(VI)]: 10^{-6} M (Osman et al. 2103a)

These complexes are of great environmental concern, especially the aquatic species $Ca_2UO_2(CO_3)_3$, which was first described and validated by Bernhard et al. (1996; 2001). It was extensively used by many authors (Zheng et al. 2003; Kelly et al. 2007; Prat et al. 2009; Nair and Merkel, 2011; Osman et al. 2013a) to understand the environmental behavior and mobility of uranium under natural conditions.

$$2Ca^{2+} + UO_2^{2+} + 3CO_3^{2-} = Ca_2UO_2(CO_3)_{3(aq)};\ \log\beta^\circ = 30.79\pm0.24 \qquad (2-5)$$

Despite the importance of these ternary complexes, they are not included in the most recent thermodynamic data of uranium that is compiled by Guillaumont et al. (2003).

An important inorganic ligand is the phosphate, which plays a role not only in environmental systems but also in biological fluids (urine, saliva, blood), and can form several complex species with uranium(VI) in aquatic solutions (Scapolan et al. 1998a). Depending on uranium and phosphate concentrations, the complexes $UO_2H_2PO_4^+$, $UO_2(H_2PO_4)_{2(aq)}$, $UO_2(HPO_4)_{(aq)}$, and $UO_2PO_4^-$ can be found in a pH range 1 – 8 (Bonhoure et al. 2007; Brendler et al. 1996). Most of the work on the $UO_2^{2+}/PO_4^{3-}/H_2O$ systems has been done at relatively low concentrations to avoid precipitation of uranyl phosphate. The homologous element Arsenic is expected to form complexes in the same manner. Three of these complexes, namely $UO_2H_2AsO_4^+$, $UO_2(H_2AsO_4)_{2(aq)}$, and $UO_2(HAsO_4)_{(aq)}$, have been detected by time-resolved fluorescence spectroscopy in the pH range 1 – 3 (Rutsch et al. 1999), but the formation constants are somewhat smaller than the analogue phosphate species.

In acidic sulfate-rich water, such as seepage waters from uranium mines after in situ leaching by sulfuric acid, uranyl ion forms binary sulfato complexes: $UO_2SO_{4(aq)}$, $UO_2(SO_4)_2^{2-}$, $UO_2(SO_4)_3^{4-}$ (Geipel et al. 1996). However, increasing in pH and/or uranium concentration leads to several ternary sulfato complexes with hydroxo bridging such as; $UO_2(OH)SO_4^-$, $UO_2(OH)(SO_4)_2^{3-}$, $(UO_2)_2(OH)_2(SO_4)_2^{2-}$, and $(UO_2)_3(OH)_4(SO_4)_3^{4-}$ (Moll et al. 2000).

Nitrate does not complex with uranium(VI) and has no role in the presence of other inorganic ligands in environmentally relevant waters. But at highly concentrated salt solutions (I = 7.0 M), the complexation becomes remarkable as the $UO_2NO_3^+$ is formed (Choppin and Du, 1992). However, in a nitric acid medium uranium(VI) coordinates with two nitrate groups to produce $UO_2(NO_3)_2.xH_2O$ (x = 2, 3, 6); x value depends on acid concentration. The valuable property of this complex is its high solubility in both water and organic solvent, which is an important feature in the processing of nuclear waste (Cotton 2006).

Chloride, which is a common ion that presents in all natural waters, forms very weak complexes with uranium(VI). Thermodynamic calculations manifest only two binding forms: UO_2Cl^+ and $UO_2Cl_2^0$, summarized in the NEA thermodynamic database as log β_1^0 = 0.17±0.02 and log β_2^0 = -1.1±0.02, respectively (Guillaumont et al. 2003). With increasing chloride concentration up to 9.0 M, higher complexes were observed: $UO_2Cl_3^-$, $UO_2Cl_4^{2-}$, $UO_2Cl_5^{3-}$ (Hennig et al. 2005). However, their formation constants have not been reported in the literature so far.

In general, the relative strength of the uranyl ion (hard acid) complexation towards inorganic ligands (hard bases) decreases in the following order: $CO_3^{2-} > OH^- > F^-, HPO_4^{2-} > SO_4^{2-} > Cl^-, NO_3^-$ (Langmuir 1997).

2.2.3 U(VI) species with organic ligands

Interaction of uranium(VI) with organic molecules commonly present in natural environments is of great concern since it modifies its solubility and toxicity. For instance, organic acids such as citrate (Cit) and oxalate (Ox), which play a major role in the metabolism activity of many organisms, can be bound to uranium(VI): UO_2Cit^-, $(UO_2)_2(Cit)_2^{2-}$, UO_2Ox, and $UO_2(Ox)_2^{2-}$ (Guenther et al. 2011; Rajan and Martell, 1965). Another common organic substance that influences the behavior of uranium(VI) in nature is humic substances (HS), including *viz*, humic acid and fulvic acid. These are complex macromolecules with reactive functional groups and carrying non-specific charges, which influence their properties, particularly solubility. The reactive functionalities in humic substances are mostly represented by carboxylate and phenolic groups. Humic substances play an important role in organic-rich fresh-surface waters with low hardness and alkalinity (< 40 mg $CaCO_3$/L), where they form dissolved species with uranyl ion (Markich 2002). The formation of uranyl-HS complexes is pH dependent; the binding affinity is increasing up to about pH 6 due to the increase in deprotonation of the carboxylate (COOH) functionalties, and then decreases at higher pH (7 – 9) where carbonate complexes are preferred. In order to simulate speciation behavior of uranium(VI) with such large polyelectrolyte molecule, simple model ligands representative of humic substance's building blocks have been extensively investigated. Some examples include complexation of uranium(VI) with malonate (Brachmann et al. 2002), with anthranilic acid and nicotinic acid (Raditzky et al. 2010), with benzoic acid (Luetke et al. 2012).

An important group of organic ligands are those of biological interest which commonly present in bio-system or bio-molecules which are secreted by microorganisms. Although little is known about uranium transport behavior and toxicity in bio-systems, some recent data is available with the associated thermodynamic stability constants. For instance, complexation of uranyl ion with some blood proteins such as trasferrin and albumin were studied (Scapolan et al. 1997; Scapolan et al. 1998b; Montavon et al. 2009) and their respective thermodynamic formation constants were used for speciation calculations in human blood serum (Montavon et al. 2009). Interaction of uranium(VI) with some sugars (α-D-glucose-1-phosphate, fructose-6-phosphate) and phospholipid model compounds has also been reported (Koban et al. 2004; Koban and Bernhard, 2007). The biologically important energy-rich molecule ATP (Adenosine 5-triphosphate) tends to bind uranium(VI) via phosphate chain of the molecule and a 1:1 complex UO_2ATP^- was confirmed with $\log \beta_{11} = 7.44 \pm 0.44$ (Geipel et al. 2000). Moreover, complex formation of uranium(VI) with amino acids L-glycine and L-cysteine was investigated to understand uranium speciation in plants and bacteria (Guenther et al. 2007). A recent study by Osman et al. (2013b) focused on complex behavior of uranium(VI) with bio-ligands, namely, urea and uric acid which are commonly present in human biological fluids (urine, sweat, serum). The data from that study was included in the equation model for a better understanding and/or accurate calculations of the intricate uranium(VI) speciation in human bio-fluids.

2.3 Uranium in natural waters

The exact knowledge of uranium concentration in natural waters is a basic requirement for calculation and spectroscopic determination of its binding forms. Uranium is introduced to natural waters by weathering of U-bearing rocks (Palmer and Edmond, 1993), and is considered as a common element that always present in surface water and groundwater with extremely broad range of concentrations vary from few ng/L to several mg/L depending on the geological formation. Much data is available about uranium concentrations in different natural waters; some values were selected and presented in Table (2-2). In this context the highly extreme values of uranium (up to 12.4 mg/L) in Finnish groundwater and in other Nordic countries should be mentioned (Karpas et al. 2005; Kurttio et al. 2005; Frengstad et al. 2000; NCRP 1999). Effect from human activities (mining, phosphate fertilizers, energy production, etc.) can add more uranium to both surface and groundwater and eventually drinking water supplies. Since U(IV) is sparingly soluble in oxic natural waters, uranium mostly occurs in hexavalent state. However, it can exist as U(IV) and/or U(VI) in anoxic fresh surface waters as well as in deep groundwater aquifers where dissolved oxygen is too low.

2.3.1 Guidelines limit for drinking water

Guidelines for uranium in drinking water worldwide show a significant variation in the recommended limits. Based on a nephrotoxicity study of uranium where a short-term high dose of uranium was given to rats and rabbits via drinking water (Gilman et al. 1998a; 1998b); uranium limits were established by many national and international health authorities. For instance, the World Health Organization (WHO) derived provisional guideline values using a Tolerable Daily Intake (TDI) approach. The TDI value for U is 0.6 μg/Kg of body weight per day was defined since 1998 (WHO 1998) and remained unchanged. Based on this value and assuming two liters daily total water consumption for a 60-Kg adult, the guideline limits for U were set. According to the continuous revision of the allocation of the TDI to drinking water, the WHO changed its provisional guideline values many times. In January 2003, WHO had changed the limit from 2 µg/L to 9 µg/L after changing the allocation of TDI to drinking water from 10% to 50%, respectively (WHO 2003). 18 months later, the value was raised to 15 µg/L based on 80% allocation of the TDI to drinking water assuming low U intake via food (WHO 2004). More recently, a new provisional guideline value of 30 µg/L was set substituting all the aforementioned values (WHO 2011). "*The provisional guideline value of 30 µg/l, which is derived from new epidemiological studies on populations exposed to high uranium concentrations, replaces the previous value derived from experimental animal studies and designated as provisional on the basis of uncertainties regarding the toxicology and epidemiology of uranium as well as difficulties concerning its technical achievability in smaller supplies. It is noted that studies on human populations, when available and of good quality, are the preferred source of health-related information to be used in deriving guideline values.*" (WHO 2011). The original WHO threshold limits of 2 µg/L was recently proposed by the Bundesanstalt für Risikobewertung (Federal Institute for Risk Assessment) in Germany for preparing infant-formula (BfR/BfS 2007), and lately a 10 µg/L for adults was legislated by the Bundesministerium für Gesundheit (German Fedral Ministry of Health) (Bundesgesetzblatt 2011). Other national authorities such as the US- Environmental Protection Agency and Health Canada recommended that uranium concentration in drinking water should not exceed 30 µg/L and 20 µg/L, respectively (US-EPA 2000; Health Canada 2009). Refer to Annex (I) on how to determine the guideline limits according to TDI (Weir 2004).

2.4 Human exposure

Uranium ubiquitously exists in our environment and is found in different levels in both abiotic and biotic systems. Depending on its solubility, mobility, and bioavailability, uranium can be taken from abiotic compartments (soil, water) by plants and eventually enters the food chain. Uranium has been determined in different types of foodstuff at various concentrations as shown in Table (2-2). Relatively high concentrations (1 – 15 µg U /Kg fresh weight) have been detected in some vegetables, cereal products, and some fish species, thus having a significant contribution to human exposure via food (Fisenne et al. 1987). Also natural waters originate from high radiation background areas, have relatively high U content, such as in some Norwegian groundwaters (highest median 34 µg/L, maximum value 750 µg/L) (Frengstad et al. 2000) and in some German mineral waters (mainly from Erzgebrige, Fichtelgebirge, Bavarian Forest, Black Forest, and Vogtland) (UBA 2005). Such high values in drinking water represent a major share of intake, if consumed regularly, thus contribute to the total body burden of uranium as well. Nevertheless, oral intake of U varies greatly depending upon local geology, dietry and drinking habits (Fisenne et al. 1987; Harley 1988, Pietrzak-Flis et al. 2001). The average daily intake of uranium in many countries has been analyzed in several studies which came up with values in a range between 0.9 to 1.5 µg U per day (equivalent 11 and 18 mBq ^{238}U per day) (Harley 1988). High exceptional intake levels (13 – 18 µg U per day) have also been reported in regions of uranium mining activities (Wrenn et al. 1985). Commonly, the intake of uranium via food and drinking water is the most important pathway of human exposure. In contrast inhalation pathway is negligible for the public since the mean U concentration in non-contaminated air is extremely low 0.1 ng/m^3 (UNSCEAR 2000a), but it becomes more significant in cases of accidents and occupational exposure. In a normal situation and for non-exposed subjects, the mean total mass of uranium in human body is approximately 56 µg (equivalent to 690 µBq ^{238}U) (Fisenne et al. 1988), of which 32 µg (56%) is deposited in the skeleton, 11 µg in muscle tissues, 9 µg in fats, 2 µg in blood, and less than 1 µg in lung, liver, and kidney each. The United Nations Scientific Committee on the Effects of Atomic Radiation (UNSCEAR 2000b) has reported an estimate of a total annual intake by human body for adults is 460 µg via ingestion of food and water and 0.59 µg from inhalation.

Table 2-2. Uranium content in selected foodstuff and natural waters

Type of water	**Typical range of U content**	**References**
Surface water	0.03 – 2.1 µg/L	(WHO 2001)
Groundwater	< 0.1 – 40µg/L	(UNSCEAR 1993)
Mineral water	<0.002 – 188.8µg/L	(Sparovek et al. 2001)
Seawater	3.3 µg/L	(ATSDR 1999)
Fresh milk	0.15 ng/g	(Galletti et al. 2003)
Bread	2.44 ng/g	(Galletti et al. 2003)
Meat	1.23 ng/g	(Galletti et al. 2003)

The behavior of uranium (intake, uptake, distribution, excretion) in the body is termed as its biokinetics (Roth et al. 2001). Uranium is considered as a non-essential element since it has no known metabolic function in animals (Berlin and Rudell, 1986), but it behaves similar to other essential metals such as alkaline earth and co-precipitates with calcium in bones (Meinrath et al. 2003). Nevertheless, not all uranium is systemically available after oral intake. More than 90% of orally ingested uranium is excreted within 24 hours through feces, while only up to 2% is absorbed from the gastrointestinal tract (Priest 2001) into the circulatory system. In blood uranium exists as uranyl cation binds to anions such as carbonate, phosphate, citrate and/or to human serum proteins, e.g. transferrin and albumin (Scapolan et al. 1997; Scapolan et al. 1998b; Montavon et al. 2009). Speciation calculations proved that complexation by serum protein and carbonate controls U(VI) distribution in blood serum with the following ratios; 35% of U(VI) bound to transferrin and albumin and 65% to carbonate (Montavon et al. 2009). The systemically available fraction (2%) is administered through a two-phase removal process, i.e. the main portion (≈70%) is eliminated through urine within 24 hours, and the remaining undergoes distribution processes in different body organs, and is finally released via the renal system revealing longer half-life in the body (Taylor and Taylor, 1997). Clearance of uranium from each body organ reveals variations in the half-lives (Wrenn et al. 1985, Sontag 1986). But under the conditions of normal daily intake, the overall elimination half-life in humans was estimated to be between 180 and 360 days (Berlin and Rudell, 1986).

2.5 Health effects of uranium

In case of incorporation, uranium may pose radiological and/or chemical risk to health. The ongoing discussion in Germany and the European Union (EU) on the maximum uranium levels in drinking water reflects the increasing awareness of uranium as a public health problem. Nevertheless, the health consequences are dependent not only upon the level and duration of exposure, but mainly upon the physical and chemical forms of uranium (WHO 2001). Like other heavy metals (e.g. Pb, W, Cd, and Hg), uranium induces chemical toxicity and causes negative impacts on tissues and organs. Kidneys are the most susceptible body organ for toxic heavy metals, including uranium. Nephritis has been reported as the major chemically induced effect in humans due to uranium incorporation (Bowman and Foulkes, 1970; Hursh and Spoor, 1973; Leggett 1989). As in the case of other toxic heavy metals, exposure to elevated uranium concentrations reduces the glumerular filtration, the tubular secretion of organic anions, and the reabsorption of filtered glucose and amino acids in the proximal tubules (Bowman and Foulkes, 1970). Studies of overexposure in experimental animals (rats and rabbits) showed a chronic nephrotoxicity after a short-term high dose of uranium (ca. 50 – 90 μg/Kg body weight per day) administered via drinking water (Gilman et al. 1998a; 1998b). Moreover, human studies from Canada and Finland also correlated high uranium intake via drinking water and nephrotoxicity (Zamora et al. 1998; Weir 2004; Kurttio et al. 2006).

On the other hand, radiological impacts of uranium on human body are attributed to its properties as a radio-element that emits the highly energetic α-particles which cause adverse effects on the living tissues. Such radiological impact of uranium is governed by its isotopic ratio, e.g. enriched uranium reveals a high radio-toxic effect in comparison to depleted uranium. Due to its weak radioactivity (low specific activity and long half-lives), still many authors believed that the radio-toxicity from natural uranium is too low to cause health hazards (Meinrath et al. 2003; Prat et al. 2010; Vicente-Vicente et al. 2010).

3. Fluorescence Spectroscopy of Uranium(VI)

The uranyl ion is one of the most interesting inorganic chemical species that emits luminescence not only in solutions, but also in most of its solid minerals (Meinrath 1998). This unique character of uranium compounds has drawn much interest, and as a consequence two important phenomena were discovered; the discovery of Stokes shift in 1852 and the discovery of radioactivity by Henry Becquerel in 1896 (Meinrath 1998). Lately, the uranyl ion was considered as a "model case of inorganic photochemistry" (Balzani et al. 1978).

3.1 Principle of fluorescence

Basically, uranyl ion or any other substance gives luminescent light when its electrons transfer from electronically excited state to the ground state. Depending on the origin of the luminescence, four categories are distinguished: photoluminescence, electroluminescence, chemiluminescence, and bioluminescence.

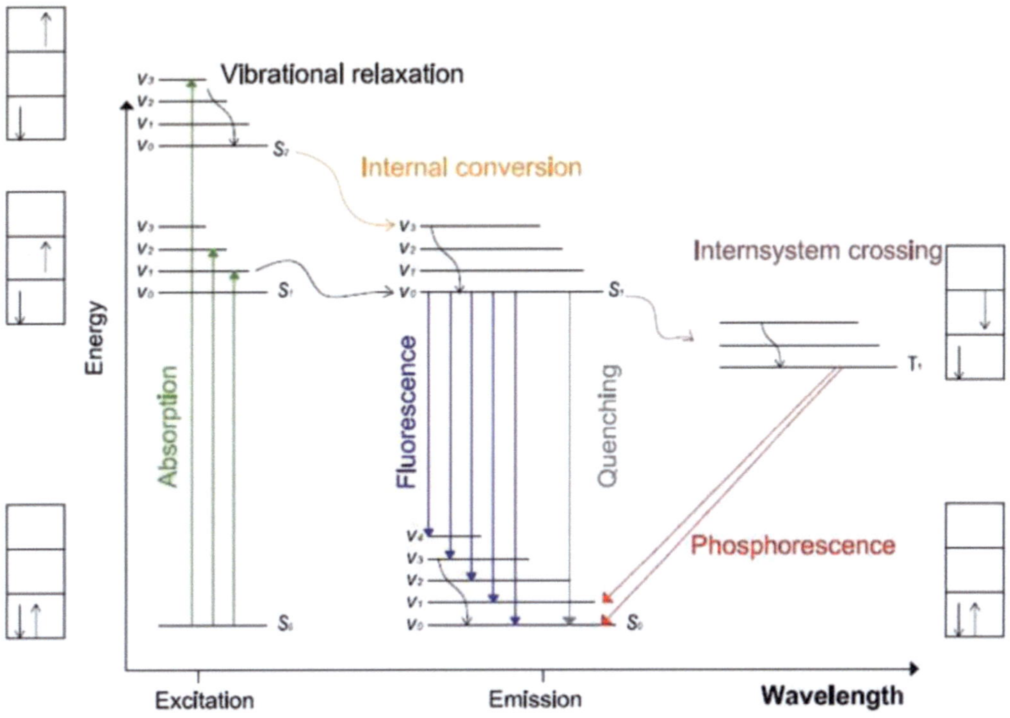

Figure 3-1. Jablonski diagram illustrating the generation and fate of a molecular excited state (Lakowicz 2006)

For the purpose of this study, details on photoluminescence (generated by absorption of photons) are reviewed. According to the nature of the excited states, photoluminescence is classified into two processes; fluorescence and phosphorescence (Lakowicz 2006). In terms of excitation and transition of electrons, all processes of molecular relaxation such as fluorescence, phosphorescence, and other non-radiative processes are normally explained by Jablonski diagram (according to the polish physicist Alexander Jablonski) (see Fig.3-1).

In a non-excited system (an atom or a molecule) all electrons exist in a ground-state orbital (S_0) at the lowest possible energy level and are paired according to Pauli Exclusion Principle. When this system absorbs energy, mainly in the form of electromagnetic radiation, electrons are excited/raised from the singlet ground-state (S_0) to a singlet excited-state (S_1 or S_2) and are located in various vibrational levels (ν_0, ν_1, ν_2, ν_3) depending on the energy of absorbed photons (Lakowicz 2006). This process is called absorption (excitation) (Fig.3-1: green arrows), it is an extremely rapid process which takes about 10^{-15} s. However, the relaxation of the electronically excited molecule to its ground state is considerably slower, ranging from 10^{-14} to several seconds (Sharma and Schulman, 1999). The electronically excited states decay via two main paths: 1) non-radiative relaxation including vibronic transition, internal conversion, and intersystem crossing; 2) radiative relaxation by emitting photons via fluorescence or phosphorescence.

The electronically excited states are not stable (vibrationally excited), so that the electron within (10^{-14} – 10^{-15} s) falls down to the lowest vibrational level of the singlet excited-state (S_2 $\nu_2 \rightarrow \nu_1 \rightarrow \nu_0$) (Sharma and Schulman, 1999). This process is known as a vibronic (combined vibrational and electronic) relaxation (Fig.3-1: black arrows) where the energy is released in the form of heat (IR quanta) or in the form of kinetic energy imparted through collision with solvent and/or solute molecules. However, if the lowest vibrational level of the upper excited state (S_2) overlaps with the highest vibrational level of the lower excited state (S_1), both excited states (S_2 and S_1) will undergo a transient thermal equilibrium allowing population of (S_1) by the electron. This vibrational coupling mechanism which permits the crossover of the excited electron from (S_2) to (S_1) is referred to as internal conversion (IC) (Fig.3-1: orange arrows), and it occurs very rapidly in about 10^{-12} s without changing the spin state (spin-allowed transition). Another radiationless process is the intersystem crossing (ISC) (Fig.3-1: dark red arrows) where there is a transition from the lowest singlet excited state (S_1) to the lower energy triplet excited state

(T_1), and it is more probable if the vibrational levels of the excited states (S_1 and T_1) overlap. Since this transition involves a conversion in the spin states, it is classically forbidden. However, from the quantum mechanics viewpoint there is a finite probability that a change of spin can occur (Sharma and Schulman, 1999). Comparing to the spin-allowed process, the probability is much lower (10^6 fold) in the case of spin-forbidden process reflecting a longer lifetime (10^{-8} s) which is almost the mean lifetime of a fluorescent species. Accordingly, intersystem crossing competes with fluorescence to deactivate the lowest singlet excited state (S_1).

Alternative to the previously mentioned modes of relaxation (non-radiative paths), the electronically excited molecule can deactivate its energy by emission of radiation (radiative paths) via either fluorescence or phosphorescence depending on the nature of the excited state (Lakowicz 2006). The excited electron may remain in the lowest vibrational level of the lowest singlet excited state (S_1) for 10^{-10} – 10^{-7} s and then drops down to one of the several vibrational levels of the singlet ground state (S_0) (Fig.3-1: blue arrows) accompanying by a rapid emission of fluorescence/photons (Sharma and Schulman 1999). The fluorescence is a rapid process with a typical lifetime in the order of 10^{-8} s. Worth mentioning that the transition is spin-allowed as the electron in the singlet excited state (S_1) is paired (by opposite spin) to the other electron in the singlet ground state (S_0). On the other hand, the phosphorescence or the radiative transition from the lowest triplet excited state (T_1) to the singlet ground state (S_0) (Fig.3-1: red arrows) is spin-forbidden (low probability of transition) revealing longer lifetime than fluorescence, typically milliseconds to several seconds (Skoog and Leary, 1996). For this reason, fluorescence is more preferred than phosphorescence for some chemical molecules. Since some energy is lost in the form of heat through vibronic relaxation and/or internal conversion before releasing the rest of energy as photons, the emission energy is typically less than that of absorption. As a result, the fluorescence spectrum typically occurs at lower energy (longer wavelength) comparing to the absorption spectrum (Fig.3-2). This shift is known as a Stokes shift after the physicist G. G. Stokes who was first observed this phenomenon as early as 1852 (Albani 2008).

Important characteristics of a fluorescent molecule (fluorophore) are: spectrum, fluorescence lifetime (τ), and quantum yield. The lifetime of the excited state is defined as the mean time that species resides in the excited state (S_1) prior to return to the ground state (S_0) (Lakowicz 2006). Mathematically, its definition is based on all processes (radiative and non-radiative) that participate in the fluorophore de-excitation (Albani 2008). The lifetime is temperature dependent and influenced by the surrounding environment of the fluorophore. The quantum yield, without complications of any mathematical formulae, is defined as the number of emitted photons relative to the numbers of absorbed photons (Lakowicz 2006).

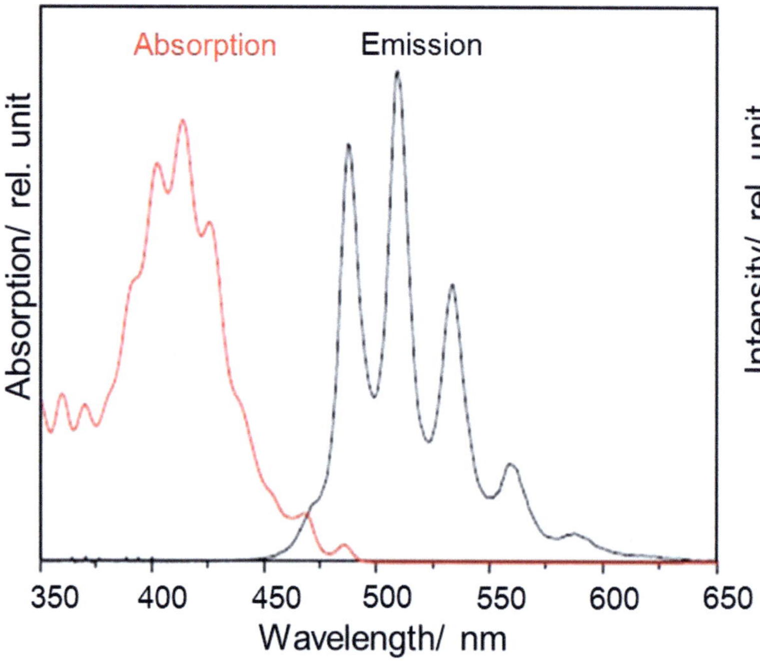

Figure 3-2. Stokes shift: The "mirror image" relationship between absorption and emission spectra in uranyl perchlorate solution

Both the emission bands (spectrum) and the lifetime are substance-specific and can be used for characterization of many different kinds of luminescent compounds such as; 1) organic species: aromatic hydrocarbons, fluorescein, rhodamines, coumarins, etc. 2) inorganic species: many actinide ions (e.g., uranyl ion UO_2^{2+}) and lanthanide ions (e.g. Eu^{3+}, Tb^{3+}), doped glasses (e.g. with Nd, Mn, Ce, Sn, Cu, Ag), crystals (ZnS, CdS, ZnSe, CdSe), etc (Valeur 2001).

The luminescence may be decreased in quantum yield or even completely vanish due to interaction of either the ground and/or excited state of the luminescent species with other species in the solution (the quencher), and this process is refer to as luminescence quenching (Sharma and Schulman, 1999). Quench effect mainly occurs via two different mechanisms; dynamic and/or static quenching, both have been extensively used in the current research. In the dynamic or collisional quenching, the interaction/collision between the quencher and the fluorophore takes place after excitation and within the lifetime of the excited state. It is also known as a diffusional quenching since it is governed by the diffusional properties of both the quencher and the luminescent molecules; which are not chemically changed in this process. The lifetime of the excited state (τ) and the concentration of the quencher species [Q] are the main limiting factors for the collisional quenching, and the decrease in the luminescence intensity can be described by the well-known Stern-Volmer equation:

$$I_0/I = 1 + K[\mathrm{Q}] = 1 + k_q \tau_0 \, [\mathrm{Q}] \qquad (3-1)$$

In this expression I_0 is the fluorescence intensity without quenching, I is the fluorescence intensity after quenching, K is the Stern-Volmer quenching constant, k_q is the bimolecular quenching constant, and τ_0 is the unquenched lifetime (intrinsic).

Since (I_0/I) is proportional to (τ_0/τ), equation (3-1) can be re-written as:

$$\tau_0/\tau = 1 + K[Q] = 1 + k_q \tau_0 \, [\mathrm{Q}] \qquad (3-2)$$

Thus the intrinsic lifetime (τ_0) can then be calculated from the following formula:

$$\tau_0 = \frac{1}{\frac{1}{\tau} - kq \cdot [\mathrm{Q}]} \qquad (3-3)$$

General examples of collisional quenchers include oxygen, halogen, amines, and electron-deficient molecules (Lakowicz 2006). In the current study more interest has been given to molecules/species that quench uranyl luminescence, such species include chloride ions, iron, manganese as well as some environmentally and biologically relevant organic molecules.

The other kind of the quenching mechanism (known as a static) does not rely on diffusion or undergo a molecular collision, but mainly implies formation of a ground state non-fluorescent complex (Valeur 2001). If we simply assume formation of a non-fluorescent 1:1 complex (MQ) between a fluorescing molecule (M) and a quencher (Q); the equilibrium reaction can be represented by the following equation:

$$\mathrm{M} + \mathrm{Q} = \mathrm{MQ} \qquad (3-4)$$

With a formation constant, *Ks*:

$$Ks = \frac{[MQ]}{[M][Q]} \qquad (3-5)$$

The mass balance on M can be given by:

$$[M_0] = [M] + [MQ] \qquad (3-6)$$

By combining Eq.(3-5) and Eq.(3-6);

$$\frac{[M]}{[M_0]} = \frac{1}{1 + Ks\,[Q]} \qquad (3-7)$$

Similarly, as in Eq.(3-3) the fluorescence intensity is proportional to the concentrations (this is only valid in diluted solutions) then Eq.(3-7) can be re-written as:

$$I/I_0 = \frac{1}{1 + Ks\,[Q]} \qquad (3-8)$$

Both quenching mechanisms yield a linear relationship when the term [(I_0/I)-1] is plotted against [Q] (Stern-Volmer plot). However, no change is observed for lifetime in the case of static quenching, whereas a decrease is clear in the case of dynamic quenching as the ratio (I_0/I) is proportional to (τ_0/τ).

3.2 Fluorescence properties of uranium

The most common oxidation states of uranium (+IV, +V, +VI) give luminescence when excited with the suitable wavelength (see Table 3-1) revealing different emission maxima (Fig.3-3). Luminescence of hexavalent uranium in solutions and/or solids remains unsettled weather it is fluorescence or phosphorescence. Brina and Miller (1992) believed that the long lifetime (a few hundred microseconds) of uranyl ion in a phosphoric acid medium at room temperature is phosphorescence and accordingly they managed to detect trace amount of uranium by laser-induced kinetic phosphorimetry. A few years later, Meinrath (1998) stated that the understanding of electronic structure of uranium(VI) indicates that the luminescence of uranyl ion is of fluorescence nature and not phosphorescence. In order to avoid such terminological confusion/contradiction, the term luminescence will be used throughout the whole text.

Table 3-1. Excitation wavelength (λ_{ex}) applicable to luminescent oxidation states of uranium along with corresponding emission maxima (λ_{em}) and luminescence lifetimes (τ)

Oxidation state	λ_{ex} (nm)	λ_{em} (nm)	(τ)	Reference
U(IV)	245	317 and others	2.7 ns	(Lehmann et al. 2009)
U(V)	255	440	1.1 µs	(Steudtner et al. 2006)
U(VI)	266, 410	470 – 588	1.6 – 2.0 µs	(Billard and Geipel, 2008)

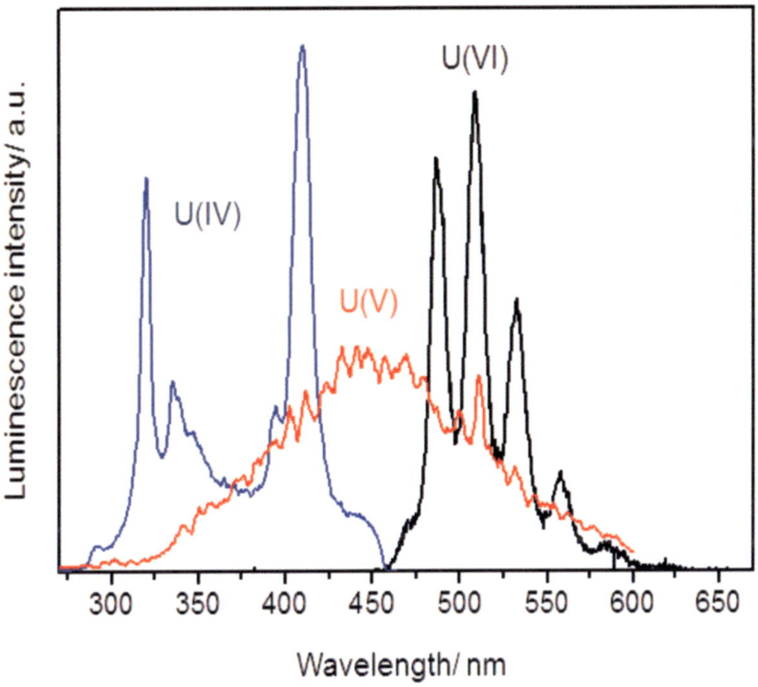

Figure 3-3. Luminescence spectra of different uranium oxidation states (IV, V, VI)

The free uranyl ion gives luminescence light mainly from two emission levels which are situated at 21270 cm^{-1} (band 470.1 nm) and 20502 cm^{-1} (band 487.8 nm) (Bell and Biggers, 1968). As depicted in figure (3-4), the emission from the higher level 21270 cm^{-1} (band 470.1 nm) represents a direct transition from an excited vibrational level of the singlet excited state (S_1) to the singlet ground state (S_0) (Meinrath 1998). However, the luminescence intensity of this emission band (470.1 nm) is relatively low and mostly not exceeding 5%, and in many cases it is non-detectable (Billard and Geipel, 2008). On the other hand, the transition from the lower emission level 20502 cm^{-1} (band 487.8 nm) reveals six vibrational levels in the ground state (S_0) corresponding to six emission bands

(Fig.3-4). The energy distance between these levels depends on the species of uranium(VI) in the solution, but still it is about 855 cm^{-1} for free uranyl ion (Bell and Biggers, 1968). Luminescence characteristics of uranyl ion and other uranium(VI) binding forms are normally determined by laser fluorescence spectroscopy (see sub-section 3.2.1).

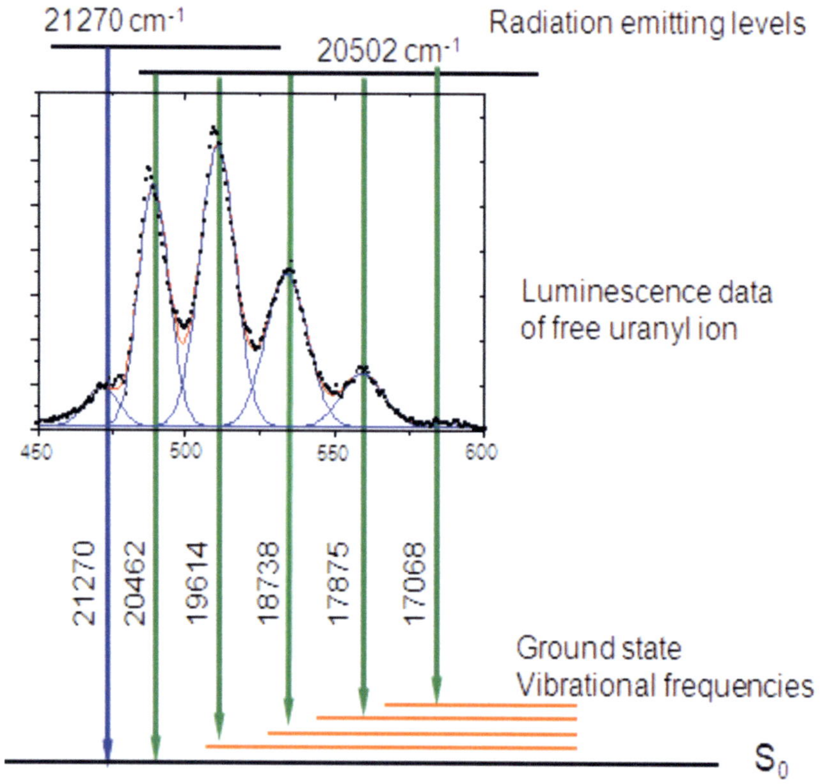

Figure 3-4. Transition of the excited electron of the aqueous uranyl ion in a perchlorate medium (Bell and Biggers, 1968)

3.2.1 U(VI) speciation by laser fluorescence spectroscopy

Several spectroscopic techniques have been developed to quantify and optimize understanding the intricate chemistry of uranium in the environment. Nevertheless, not all techniques are suitable for environmentally relevant concentrations of uranium (Fig.3-5).

A unique, rapid, and sensitive method for *direct* uranium speciation in the framework of nuclear, environmental, and biological studies is the time-resolved laser-induced fluorescence spectroscopy (TRLFS). This technique is based on a laser excitation of the uranyl ion by a suitable wavelength, and the recorded luminescent spectrum is observed by simultaneous detection of peaks positions and the luminescence lifetime.

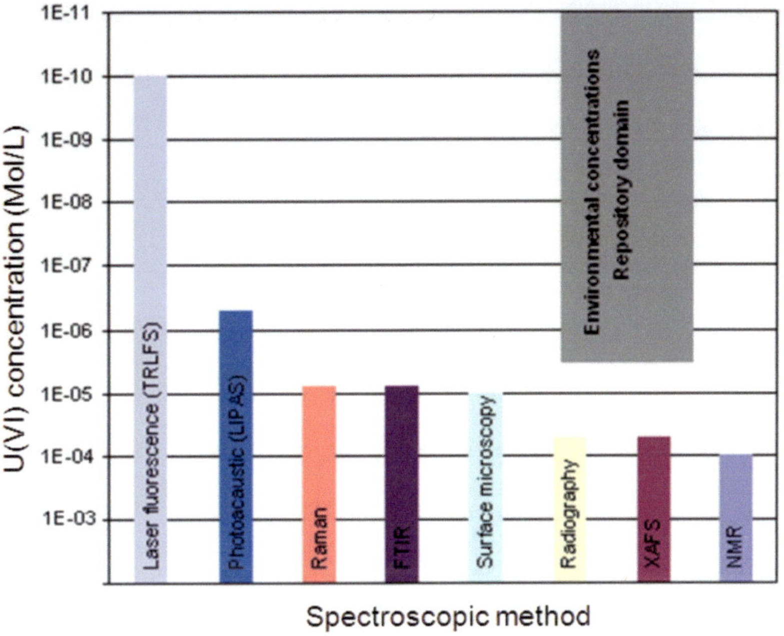

Figure 3-5. Sensitivity of the TRLFS in comparison with other spectroscopic methods (diagram was taken from Runde 2000)

By comparing these luminescence characteristics with corresponding spectra of known uranium compounds it is possible to identify the unknown species. Unlike traditional excitation sources (e.g. lamps), laser sources have the major advantage of generating a highly monochromatic light, thus samples can be excited by a defined wavelength. Besides that, the energetic-rich laser radiation can excite trace amount of the analyte (e.g. UO_2^{2+}) in the sample making the detection limit very low. Moreover, very fast processes can be investigated since laser sources can generate very short pulses (range of μs to fs). Rather than other oxidation states, uranium(VI) can be excited by many wavelengths as shown in Table (3-1). For instance, the 266 nm has higher excitation energy and can excite all uranyl compounds non-specifically. Some environmental samples have high contents of natural organic matter which can also absorb this wavelength making the excitation of uranium(VI) very week. Hence, using other wavelength is imperative, e.g. the 410 nm which is not absorbed by organics can also excite uranium(VI), but it is more species dependent. The TRLFS spectrum of a free uranyl ion has a unique shape of five fingers (characteristic fingerprint). The associated luminescence lifetime at pH 2, 25°C, and 0.1 M $NaClO_4$ solution was reported in the range 1.6 – 2.0 μs (Billard and Geipel, 2008). However, the luminescence lifetime of free uranyl ion is strongly influenced by the surrounding environment like temperature, medium, presence of quenchers (Meinrath 1998). Typical luminescence characteristics of the free uranyl ion is illustrated in figure

(3-6) and detailed in Table (3-2). Details about TRLFS setup and measurement are given in the section "Experimental Details".

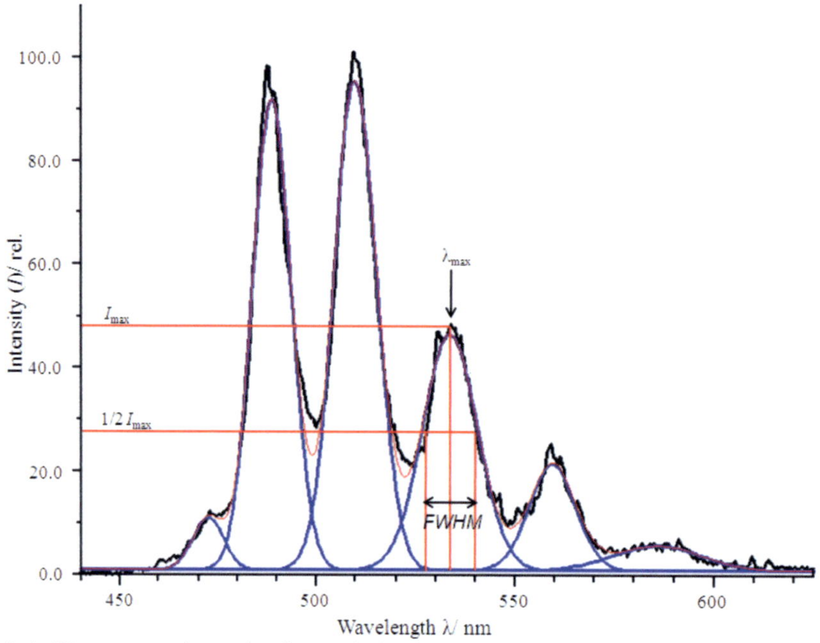

Figure 3-6. A 2D spectrum shows the fingerprint/peaks characteristics of the free uranyl ion (pH 2, $I_{0.1M}$, 25°C) when excited by a wavelength λ_{ex} = 266 nm (own measurement)

Table 3-2. Detailed luminescence data of UO_2^{2+} solution (pH:2, I:0.1M, 25°C), λ_{ex} = 266 nm

• **Lifetime (µs)**			
- 1.95±0.05			
• **Emission λ_{max} (nm)**	**FWHM* (nm)**	**Area** (%)**	**Intensity I_{max}**
- 472.0±0.3	10.1±0.5		10.7±0.3
- 488.7±0.2	11.6±0.1	30.4±0.3	93.5±1.8
- 509.7±0.2	12.5±0.2	35.1±0.2	100
- 533.5±0.2	16.5±0.3	22.2±0.3	47.9±1.0
- 559.4±0.2	15.1±0.8	8.4±0.4	19.8±1.0
• **Band spacing (cm^{-1})**			
- 846±8			

*Full Width at Half Maxima, ** Area under the peak

A great advantage of the TRLFS is the possibility of analysis of multi-component systems, i.e. the time resolved measurements make the separation into individual components possible. Hence, various uranium(VI) species/complexes can be characterized by TRLFS via providing; 1) spectral information (peaks shift and/or intensity variation), 2) temporal information (lifetime resolution). These properties are of great importance with respect to the application of TRLFS to uranium(VI) speciation in natural samples (environmental and biological). In natural systems, where organics are highly present, uranium(VI) can easily be featured by its longer luminescence lifetime (micro second range) comparing to fluorescent organics (femto second range).

3.3 Chemical quenching of uranyl luminescence

Luminescence of uranyl complexes in natural samples (environmental and/or biological) are strongly influenced by presence of chemical quenchers (organic and/or inorganic). Various mechanisms have been suggested to elucidate the quenching process. One mechanism includes intermolecular electron transfer from the quencher to the excited uranyl ions (Burrows et al. 1976; Burrows and Pedrosa de Jesus, 1976; Yokoyama et al. 1976; Moriyasu et al. 1977) which may prompt a radiationless transition of the excited uranyl ion to the ground state (Matsushima et al. 1974). Quenchers representing electron transfer mechanism include inorganic ions such as some metal ions (Fe^{3+}, Mn^{2+}, Co^{2+}, Ce^{3+}, Hg_2^{2+}, Ag^{+}), NO_2^-, CNS^-, and halides (with the exception of F^-). Organic substances/quenchers undergo a different mechanism including energy transfer from the excited uranyl ions to the quencher (Kropp 1967). The uranyl ion in its electronically excited state, denoted as $^*UO_2^{2+}$, is very reactive and is expected to be a powerful oxidizing agent, and the energy of the excited electron corresponds to a redox potential of about 2.6 V (Balzani et al. 1978). Thus, photochemical reactions/quenching can be induced by transfer the energy of ($^*UO_2^{2+}$) to organic substances present in the medium, making difficulties when characterizing U(VI) species using TRLFS.

Beside natural organic matter, the water molecules and carbonate/bicarbonate ions represent important quenchers in the framework of U(VI) speciation in aqueous natural samples using TRLFS. Quenching process by water molecules was extensively discussed (Moriyasu et al. 1977; Marcantonatos 1980; Formosinho et al. 1984; Meinrath 1998) and believed to happen via electron transfer (Eq.3-9) and a hydrogen abstraction reaction

(Eq.3-10), which can also be counted as a kind of electron transfer accompanied by proton transfer.

$$^{*}UO_2^{2+} + H_2O = UO_2^{+} + H^{+} + OH^{o} + e^{-} \qquad (3-9)$$

$$^{*}UO_2^{2+} + H_2O = UO_2H^{2+} + OH^{o} + e^{-} \qquad (3-10)$$

The standard oxidation-reduction potential (Eh^{o}) of hydrogen abstraction is pH dependent (Moriyasu et al. 1977) as illustrated in equation (3-11);

$$OH^{o} + H^{+} + e^{-} = H_2O\;; \qquad Eh^{o} = 2.82 - 0.059\,\mathrm{pH} \qquad (3-11)$$

Thus, quenching by water molecules is in prospect to decrease with decreasing pH value, this was supported by experiments and further justified by consideration of molecular orbitals (Moriyasu et al. 1977). Decrease in the pH increases the affinity of non-bonding orbitals in the oxygen atom of the water molecule to highly coordinate with the proton and forming hydronium ions (H_3O^{+}). As a consequence the abstraction of hydrogen from H_3O^{+} by the excited uranyl ion should be more difficult or slower than from H_2O. This explains the dependency of luminescence lifetime on pH as well as the bi-exponential curves at pH > 3 (Moriyasu et al. 1977). Marcantonatos (1980) demonstrated the mechanism of hydrogen abstraction through the formation of the exciplex $^{*}[(UO_2)_2H]^{4+}$ according to the following equations:

$$^{*}UO_2^{2+} + H_2O = {}^{*}UO_2H^{2+} + OH^{o} \qquad (3-12)$$

$$^{*}UO_2H^{2+} + UO_2^{2+} = {}^{*}[(UO_2)_2H]^{4+} \qquad (3-13)$$

A common quencher of uranyl luminescence in nature is the carbonate/bicarbonate system. The quenching mechanism was proposed via formation of a carbonate radical CO_3^{o-} ($^{o:}$ unpaired electrons) after the experiment of flash light photolysis of a solution containing 2 mM U(VI) and 10 mM $NaHCO_3$ (Balzani et al. 1978).

$$UO_2^{2+} + HCO_3^{-} + h\nu = UO_2^{+} + CO_3^{o-} + H^{+} \qquad (3-14)$$

The U(V) ion which is produced in equation (3-9) and equation (3-14) is unstable in aqueous medium and disproportionates according to:

$$2UO_2^{+} + 4H^{+} = UO_2^{2+} + U^{4+} + 2H_2O;\;\; E^{o} = 0.19\ \mathrm{V} \qquad (3-15)$$

The kinetic of the above reaction is slow including break up of covalent bonds. Therefore the photochemical oxidation of U(V) by the produced radicals (OH^{o}, CO_3^{o-}) is possible, and this explained the absence of U(IV) in the carbonate/water system after irradiation (Meinrath 1998). Due to HCO_3^-/CO_3^{2-} quenching, higher carbonate species $UO_2(CO_3)_2^{2-}$ and $UO_2(CO_3)_3^{4-}$ were known to be non- luminescent at ambient temperature (Kato et al. 1994). However, a recent TRLFS measurement of a solution containing 2 mM U(VI) and 500 mM $NaHCO_3$ at pH 8.2 and at 293 K confirmed the first luminescence spectrum of tris-carbonato uranyl ($UO_2(CO_3)_3^{4-}$) (Götz et al. 2011).

3.4 Determination of the complex stability constant from spectroscopic data

In the case of fluorescence spectroscopic analysis of complex formation reactions; two different phenomena can be observed in the resulted spectra:

- Bands shifts and intensity increase when compared to the pure metal or ligand spectrum.
- A decrease in the fluorescence intensity of the fluorophore by addition of another substance (fluorescence quenching), i.e. the formed complex is non-fluorescent.

In both cases, the stability constant of the formed complex (Log β) can be calculated by slope analysis. In this work the slope analysis has been applied to the special case of fluorescence quenching (intensity decrease) upon complexation, and the process can be described by the law of mass action. However, when both static and dynamic quenching exists in a complex formation reaction, the dynamic component (decrease in the lifetime) of the fluorophore should be related as a function of the concentration of the quencher. On the other hand, the fluorescence intensities without the contribution of the dynamic quenching can be evaluated by the Stern-Volmer relation (Stern and Volmer, 1919) that is previously shown in (Eq.3-1):

$$I_0/I = 1 + K[\mathrm{Q}] \qquad (3-1)$$

Where I_0 is the fluorescence intensity without quenching, I is the fluorescence intensity after quenching, K is the Stern-Volmer quenching constant. When the term (I_0/I)-1 is plotted versus the concentration of the quencher, a straight line is yielded and then the slope (K) directly represents the association constants of the respective complex.

If we simply assume formation of a non-fluorescent complex (ML) between a fluorescent metal (M) and a complexing ligand (quencher) (L); the complexation reaction can be represented by the following equation:

$$M + LH \leftrightarrow ML + H^{+} \qquad (3-16)$$

Applying the mass action law with the above complex reaction, the formation constant (β) can be given by:

$$\beta = \frac{[H^{+}][ML]}{[M][LH]} \qquad (3-17)$$

Equation (3-18) can be converted to the linear form by taking logarithm values for both sides:

$$Log\frac{[ML]}{[M]} - pH = \text{Log [LH]} + \text{Log}\,\beta \qquad (3-18)$$

Plotting the logarithm of the concentration ratio of the complex (ML) and the free metal (M) to the logarithm of the free ligand concentration (LH) results in a straight line and the slope corresponds to the stoichiometry of the formed complex. The corresponding formation constant can be obtained from the intercept (Log β in Eq.(3-19)). Worth mentioning is that determination of the gross stability constant of the formed complex (ML) should include all stability constants of the individual components of the complex formation reactions (e.g. protonation/deprotonation of the ligand). However, sometimes the slope analysis method cannot be applied when several complexes of the type ML_n coexist. In this case, the factor analysis program SPECFIT can be used for the determination of the stability constants. Information about SPECFIT is found in the section "Experimental Details".

3.5 Fluorescence properties at low temperatures

As previously mentioned, the luminescence properties are dependent on the surrounding environment of which temperature and presence of quenchers are the most important factors. Quenchers in aqueous natural samples influence not only the luminescence yield of uranium(VI), but also the lifetime of the excited state via a collision process which results in a radiationless deactivation (dynamic quenching). Since this type of quenching depends on the diffusional properties of the molecules in the surroundings, it can then be minimized via restricting the free movement of the molecules by freezing the sample. Thus, lowering temperatures below freezing point reduces quench effects as the non-radiative transfer of energy from the excited uranyl ion decreases (Geipel 2006). With regards to that, the luminescence intensity is enhanced, the lifetime is elongated, and the spectral resolution is improved. Lotnik et al. (2003a) reported the dependence of lifetime of electronically excited state of uranyl ion on temperature in a sulphuric acid medium. They further investigated the effect of phase transition between frozen and aqueous phase (Lotnik et al. 2003b). The same behavior of temperature effect on uranyl luminescence spectra of environmentally relevant aqueous samples was observed by Wang et al. (2004). They characterized fluorescence properties of uranyl-carbonate species in highly contaminated pore waters from Hanford Vadose Zone in the United States. They used TRLFS at liquid helium temperature (6 K) and managed for the first time to define the spectral data for ($UO_2(CO_3)_2^{2-}$, $UO_2(CO_3)_3^{4-}$, and $(UO_2)_2(OH)_3CO_3^-$) which are non-fluorescent at room temperature. The spectral data are detailed in Table (4-4) in the section "Results and Discussion". Application of laser fluorescence spectroscopy coupled with cold temperature system (cryo-TRLFS) can offer a new possibility for direct determination of uranium(VI) species in natural samples. The cryo-TRLFS is the methodical basis of the current research as elaborated in the section "Experimental Details".

4. Results and Discussion

4.1 Investigation of U binding forms in environmentally relevant waters

Several water samples (mineral water, groundwater, mining water, and wastewater) have been investigated in the framework of uranium speciation at environmentally relevant concentrations.

4.1.1 U(VI) speciation in selected German mineral waters

Much focus was given to this type of water as it is a challenge to get detailed information about uranium speciation at *exceptionally low concentrations* (< 2.0 µg/L), which could be the base for further study for low dose and long term exposure. In addition, the pH range (5.5 – 7.5) is quite interesting, especially for carbonated-waters, in which the U distribution might change due to changes in pH and CO_3^{2-}/HCO_3^- content. The approach is to find out, experimentally, the uranium binding forms at such *extremely low level* and compare the findings with the uranium thermodynamic modeling results of these waters.

4.1.1.1 Water samples and analytical data

Seven mineral waters were bought from local supermarkets in Lower Saxony, Bavaria, Hesse, Thuringia, Baden-Württemberg, and the North Rhine-Westphalia States, Germany. Uranium concentrations were substantially low (0.4 – 2.0 µg/L). Based on bicarbonate content and pH value, the water samples were categorized into; 1) carbonated water (CLASSIC) with a relatively high carbonate content and a pH value between 5 and 6; and 2) non-carbonated water (STILL) with pH 6.5 – 7.5. The samples Adelholzener, Alwa, Teinacher, and Warburger Waldquelle are CLASSIC mineral waters, while samples from Extalerquelle, Nestle Aquarel, and Thüringer Waldquelle are representing STILL type (see Table 5-8 in the section "Experimental Details"). Water samples were analyzed for main inorganic constituents, total inorganic carbon (TIC), and total organic carbon (TOC) (see Table 4-1). Uranium(VI) speciation was measured by cryo-TRLFS. Details on sample preparation for laser measurements, cryo-TRLFS, and the analysis techniques are in the section "Experimental Details".

Table 4-1. Concentrations of some water quality parameters (inorganic elements) and total organic compounds

Sample ID	pH	U(VI) (µg/L)	Na^{+} (mg/L)	Mg^{2+} (mg/L)	Ca^{2+} (mg/L)	F^{-} (mg/L)	Cl^{-} (mg/L)	SO_4^{2-} (mg/L)	TIC (mg/L)	TOC (mg/L)	Electrical conductivity (µS/cm)	Ionic strength (mol/L)
Extalerquelle (STILL)	6.88	2.09	16.7	57.5	322	0.05	13.4	857	47.8	0.81	1740	0.028
Nestle-Aquarel (STILL)	7.56	0.408	64.5	31.1	109	0.46	77.5	19.1	108	0.81	974	0.016
Thüringer Waldquell (STILL)	6.88	0.931	20.2	38.3	83.9	0.70	35.3	172	51.9	<0.5	744	0.012
Adelholzener (CLASSIC)	5.48	1.42	13.1	31.3	67	0.19	19.1	26.3	386	0.88	568	0.009
Alwa (CLASSIC)	5.68	1.62	15.1	59.7	460	0.32	28.2	990	149	1.50	2210	0.036
Teinacher (CLASSIC)	5.45	1.81	71.6	50	190	1.07	14.7	25.2	213	0.75	855	0.014
Warburger Waldquell (CLASSIC)	5.78	1.66	21.7	45.9	250	1.52	15.8	495	179	0.87	1408	0.023

4.1.1.2 Speciation modeling of the mineral waters

Species distribution of uranium in mineral waters were calculated by the computer code EQ3/6 (Wolery 1992) using up-to-date and well-reviewed thermodynamic formation constants for dissolved uranyl complexes taken from the NEA-TDB (Nuclear Energy Agency-Thermochemical Database) (Guillaumont 2003). Additional data for aqueous $Ca_2UO_2(CO_3)_3$ and $CaUO_2(CO_3)_3^{2-}$ complexes were taken from Bernhard et al. (2001) and $MgUO_2(CO_3)_3^{2-}$ from Dong and Brooks (2006). Analytical water parameters that have been used in the calculations were; pH, cations (Ca^{2+}, Mg^{2+}, Na^+), anions (HCO_3^-, SO_4^{2-}, Cl^-, F^-), and U concentration. The low ionic strength of the waters (0.009 – 0.036 M) has no influence on uranyl ion speciation. Each ion/ligand in the water has its own binding behavior with uranium depending on the pH range and the ligand concentration, for instance UO_2F^+ and UO_2SO_4 exist in acidic waters (pH < 4.0) such as in mining waters. In the pH range of the investigated waters (5.5 – 7.5) carbonate ligands control uranium distribution in aqueous solutions forming different types of uranyl-carbonate species. Figure (4-1) was used to elucidate the influence of the TIC on the U(VI) distribution in solution as a function of pH, taking into consideration an equilibrium: (a) in the absence of CO_2 and, (b) with atmospheric CO_2 ($p(CO_2)$: $10^{-3.5}$ bar).

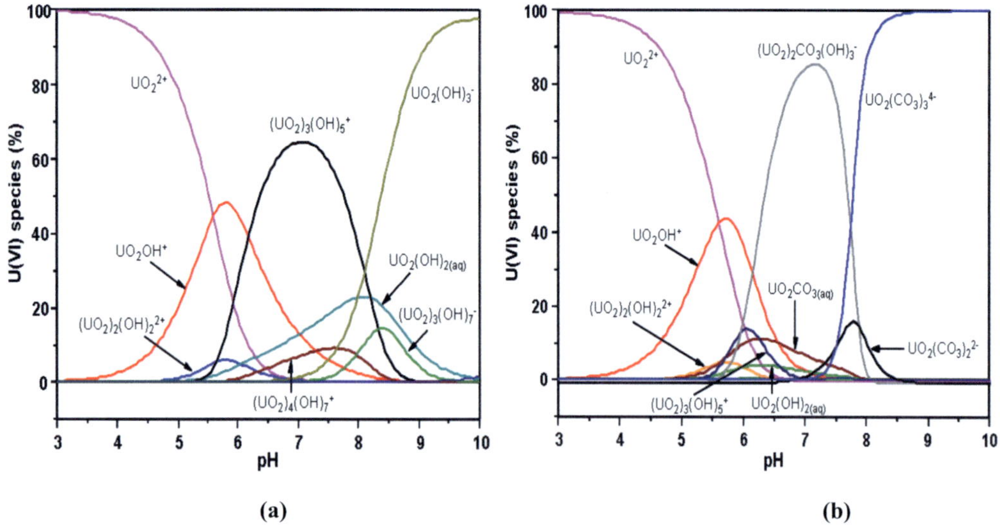

Figure 4-1. U(VI) binding forms in aquatic solutions in the **(a)** absence of CO_2 and **(b)** $p(CO_2)$: $10^{-3.5}$ bar, and [U(VI)]: 10^{-6} M.

Table 4-2. Some important formation constants for aqueous uranyl complexes used for thermodynamic modeling of mineral waters

Uranyl complex	$\log\beta_{I=0M,\ T=25°C}$	Uncertainty (±)	Source
UO_2OH^+	-5.250	0.240	Guillaumont et al. 2003
$UO_2(OH)_{2(aq)}$	-12.150	0.070	Guillaumont et al. 2003
$(UO_2)(OH)_2^{2+}$	-5.620	0.040	Guillaumont et al. 2003
$UO_2(OH)_3^-$	-20.250	0.420	Guillaumont et al. 2003
$UO_2(OH)_4^{2-}$	-32.400	0.680	Guillaumont et al. 2003
$(UO_2)_2OH^{3+}$	-2.700	1.000	Guillaumont et al. 2003
$(UO_2)_3(OH)_4^{2+}$	11.900	0.300	Guillaumont et al. 2003
$(UO_2)_3(OH)_5^+$	-15.550	0.120	Guillaumont et al. 2003
$(UO_2)_3(OH)_7^-$	-32.200	0.800	Guillaumont et al. 2003
$(UO_2)_4(OH)_7^+$	-21.900	1.000	Guillaumont et al. 2003
$UO_2(CO_3)_2^{2-}$	16.610	0.090	Guillaumont et al. 2003
$UO_2(CO_3)_3^{4-}$	21.840	0.040	Guillaumont et al. 2003
$(UO_2)_2CO_3(OH)_3^-$	-0.858	0.851	Guillaumont et al. 2003
$(UO_2)_3(CO_3)_6^{6-}$	54.000	1.000	Guillaumont et al. 2003
$UO_2CO_{3(aq)}$	9.940	0.030	Guillaumont et al. 2003
$UO_2SO_{4(aq)}$	3.150	0.020	Guillaumont et al. 2003
$UO_2(SO_4)_2^{2-}$	4.140	0.070	Guillaumont et al. 2003
$UO_2(SO_4)_3^{4-}$	3.020	0.38	Guillaumont et al. 2003
UO_2Cl^+	0.170	0.020	Guillaumont et al. 2003
$UO_2Cl_{2(aq)}$	-1.100	0.400	Guillaumont et al. 2003
UO_2F^+	5.160	0.060	Guillaumont et al. 2003
$Ca_2UO_2(CO_3)_{3(aq)}$	30.790	0.240	Bernhard et al. 2001
$CaUO_2(CO_3)_3^{2-}$	25.600	0.250	Bernhard et al. 2001
$MgUO_2(CO_3)_3^{2-}$	26.110	0.040	Dong and Brooks (2006)

However, such speciation changes significantly in the presence of cationic species (Ca^{2+} and Mg^{2+}) commonly exist in natural waters. These cations form different mobile aqueous uranyl species of which $Ca_2UO_2(CO_3)_{3(aq)}$ is the most important and dominates the speciation diagrams in all investigated mineral waters (Fig.4-2 and 4-3). For instance, at a fixed pH of 6.88 and $[HCO_3^-]$ = 4.3 mM, calcium ion controls U(VI) distribution in Thüringer STILL water (Fig.4-2(b)), where 1 mM Ca^{2+} is enough to form more than 90% of $Ca_2UO_2(CO_3)_{3(aq)}$.

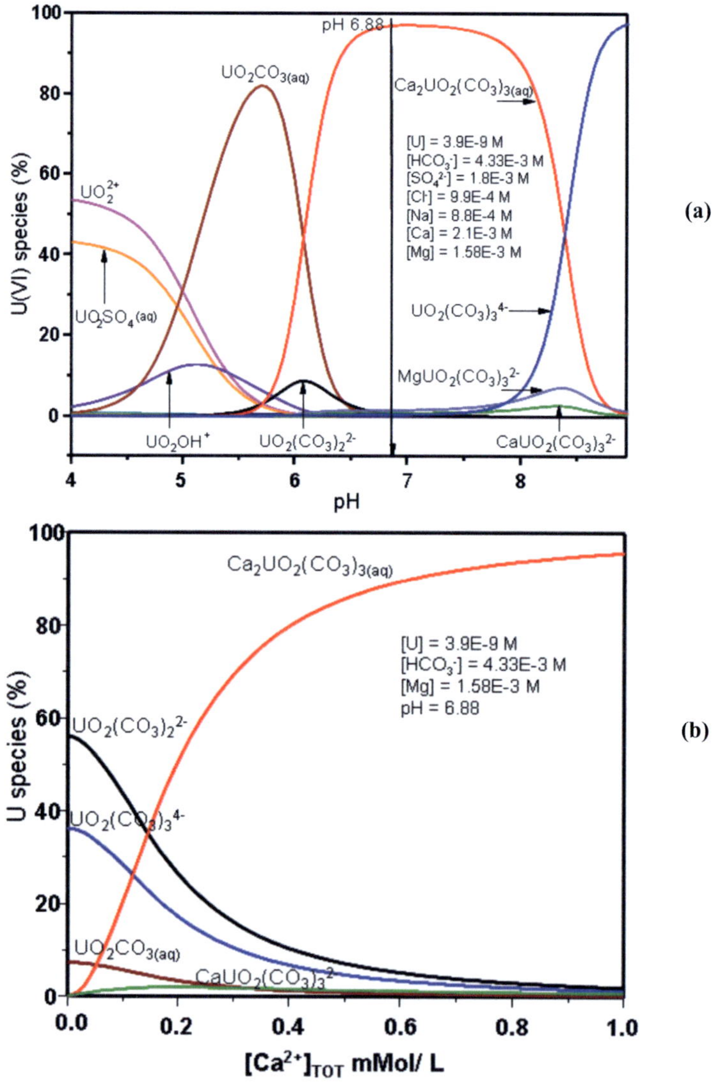

Figure 4-2. U(VI) distribution as a function of **(a)** pH and **(b)** $[Ca^{2+}]$ in Thüringer STILL water

However, the same concentration of Ca^{2+} (1 mM) yields only 50% of $Ca_2UO_2(CO_3)_{3(aq)}$ in Adelholzener CLASSIC water with a higher bicarbonate content and a fixed pH of 5.5 (Fig.4-3). Detailed results of the calculated binding forms of U in the investigated water samples are presented in Table (4-3). Both calculations and fluorescence spectroscopic measurements display $Ca_2UO_2(CO_3)_{3(aq)}$ as a dominant binding form of uranium in the studied waters.

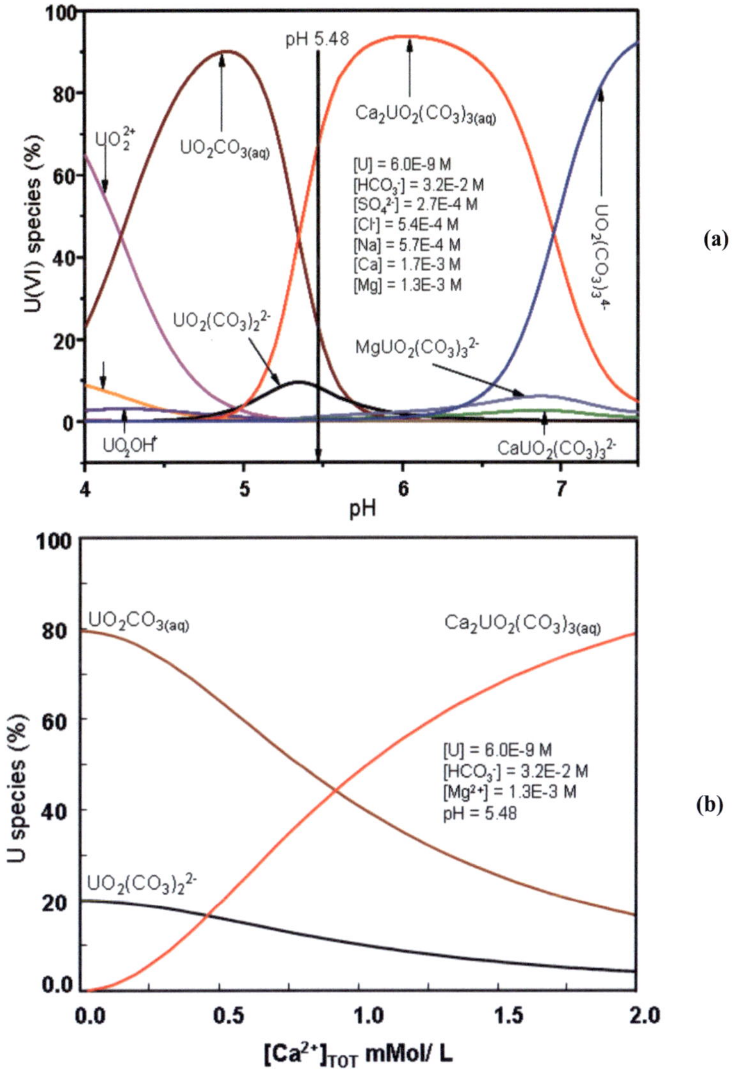

Figure 4-3. U(VI) distribution as a function of **(a)** pH and **(b)** $[Ca^{2+}]$ in Adelholzener CLASSIC water

Table 4-3. Expected uranium binding forms in the investigated mineral waters according to calculations performed using NEA-TDB and the geochemical speciation code EQ3/6

Sample ID	Uranium binding form (%)		
	$Ca_2UO_2(CO_3)_{3(aq)}$	$UO_2CO_{3(aq)}$	$UO_2(CO_3)_2^{2-}$
Extalerquelle	99.3	-	-
Nestle Aquarel	98.4	-	-
Thüringer Waldquell	97.0	-	-
Adelholzener	69.0	20.7	8.2
Alwa	95.7	2.5	1.1
Teinacher	74.5	20.1	4.3
Warburger Waldquell	96.3	1.7	1.3

4.1.1.3 Luminescence characteristics of uranium in mineral waters

Figure (4-4) shows a typical time-resolved spectrum of uranium in a mineral water sample (Extalerquelle) at 153 K. All mineral waters under investigation showed non-fluorescent properties at room temperature due to the strong quench effect of carbonate ions and water molecules in the solvation shell. Lowering the temperature minimizes the quench effect since the energy transfer from the excited uranyl ion decreases as reported by Geipel (2006). Hence, the luminescence intensity is enhanced, the lifetime is elongated, and the spectral resolution is improved, i.e. frozen mineral waters showed fluorescent species. This temperature effect on fluorescence spectra of aqueous uranyl species was first observed by Wang et al. (2004). They characterized fluorescence properties of uranyl-carbonate complexes in pore water samples from Hanford contaminated sediments using TRLFS at liquid helium temperature (6 K). Our cryogenic TRLFS system was also dedicated for such measurements using the optimized parameters that were already mentioned in the section "Experimental Details".

Referring to Table (4-4), we see that all types of waters showed very little differences in the main emission bands. Accordingly, identification of uranium species in such samples depending only on band position is quite difficult. However, it could be possible to assign the speciation in this case via lifetimes, which can be compared with lifetimes of uranyl carbonate species in model solutions as shown in Table (4-4). The luminescence decay curves were described by both mono- and bi-exponential functions for STILL and CLASSIC waters, respectively (see Annex II).

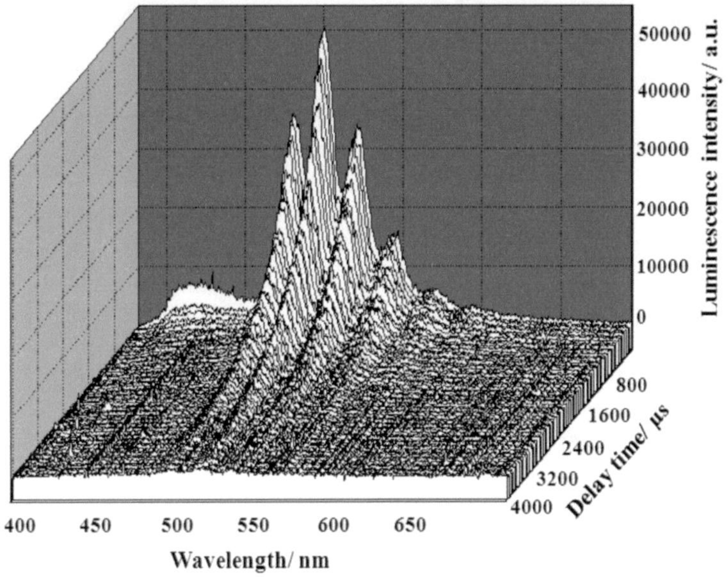

Figure 4-4. A typical time-resolved spectrum of U(VI) in a mineral water sample (water from Extalerquelle)

4.1.1.4 Uranium speciation in (STILL) water

The mono-exponential function which fit the fluorescence decay of uranyl in these waters indicated a single species which was characterized as the aqueous $Ca_2UO_2(CO_3)_3$ with a fluorescence lifetime > 1 ms. This is in consistent with the literature value as presented in Table (4-4). Moreover, we observed no spectral shift of the individual emission peaks with increasing the delay time which also proved a single species (Fig.4-5). The species $Ca_2UO_2(CO_3)_{3(aq)}$, at these low concentrations and at a pH range (6.5 – 7.5), gives no fluorescence at room temperature due to carbonate quenching. However, at relatively high concentrations of uranium and calcium it fluoresces at 465, 484, 505, and 525 nm with a fluorescence lifetime of 43 ns as reported by Bernhard et al. (1996). Prat et al. (2009) characterized the same fluorescence properties of the same species at room temperature in some Finnish groundwater samples of exceptionally high uranium content (> 600 µg/L).

Table 4-4. Luminescence characteristics of U(VI) in seven German mineral water samples at 153 K, $\lambda_{ex.}$= 266 nm in comparison with some literature data

Sample ID	**Main emission bands (nm)**	**1st lifetime (µS)**	**2nd lifetime (µS)**	**Major luminescent species**
Adelholzener	481.6, 501.5, 523.0, 545.9, 572.9	265.7±37.9	1159.2±55.7	$Ca_2UO_2(CO_3)_{3(aq)}$
Alwa	481.5, 501.6, 523.0, 545.5	258.0±43.2	1288.3±82.9	$Ca_2UO_2(CO_3)_{3(aq)}$
Teinacher	481.2, 501.5, 523.2, 545.8	382.9±1.2	1420.2±4.1	$Ca_2UO_2(CO_3)_{3(aq)}$
Warburger Waldquell	481.4, 501.5, 522.7, 545.3	347.4±38.1	1297.8±88.9	$Ca_2UO_2(CO_3)_{3(aq)}$
Extalerquelle	481.6, 501.8, 523.1, 545.5, 572.2	1013.9±17.3		$Ca_2UO_2(CO_3)_{3(aq)}$
Nestle Aquarel	481.6, 501.8, 523.2, 545.7	1061.4±54.3		$Ca_2UO_2(CO_3)_{3(aq)}$
Thüringer Waldquell	481.5, 501.0, 521.9, 544.9	1341.4±34.6		$Ca_2UO_2(CO_3)_{3(aq)}$
(Wang et al. 2004)	480.5, 501.2, 522.7, 546.0, 571.9	1282		$Ca_2UO_2(CO_3)_{3(aq)}$
(Wang et al. 2004)	479.6, 499.2, 519.9, 542.4, 565.6	883		$UO_2(CO_3)_3^{4-}$
(Wang et al. 2004)	479.0, 498.0, 519.0, 542.0, 567.0	465		$UO_2CO_{3(aq)}$
(Wang et al. 2004)	477.4, 496.4, 517.2, 539.8, 563.5	962		$UO_2(CO_3)_2^{2-}$
(Wang et al. 2004)	523.0, 542.3, 561.3	144		$(UO_2)_2(OH)_3CO_3^-$

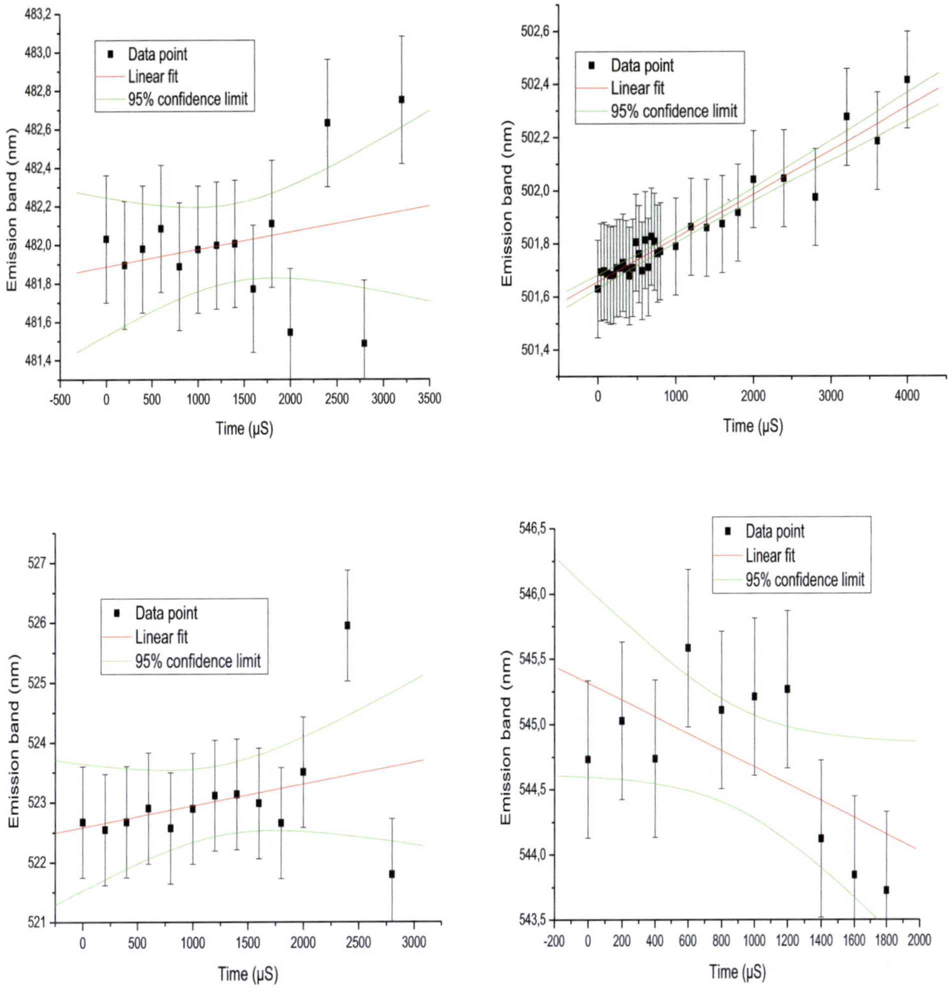

Figure 4-5. Analysis of individual emission peaks as a function of delay time in Extalerquelle as an example of STILL water

4.1.1.5 Uranium speciation in (CLASSIC) water

These waters were characterized by a pH range 5.0 – 6.0 and a high total inorganic carbon (TIC), mostly in the form of $CO_{2(aq)}$ and to lesser extend HCO_3^-. In a closed water bottle the $CO_{2(g)}$ above the water is in a dynamic chemical equilibrium with the CO_2 dissolved in the water, which in turn is in chemical equilibrium with the bicarbonate species (carbonic acid equilibrium).

$$CO_{2(g)} = CO_{2(aq)} \qquad (4-1)$$

$$CO_{2(aq)} + H_2O_{(l)} = HCO_3^-{}_{(aq)} + H^+{}_{(aq)} \qquad (4-2)$$

When the bottle is opened, the CO_2 in the gaseous form is released and the equilibria (4-1) and (4-2) are shifted to the left. Consequently a change in the water pH and bicarbonate concentrations takes place which could result in a change of uranium species distribution in the system. This created some difficulty not only in the TRLFS measurement but also with uranium speciation calculations since we do not know the new partial pressure $p(CO_2)$ in the new equilibria. We tried to take the sample inside a glove box filled with CO_2 gas, but this did not work perfectly since the $p(CO_2)$ inside the glove box was different from that inside the bottle and the equilibria were also disturbed. So, we simply took the case of the consumers/people, opening the bottle and attaining a new HCO_3^-/CO_2 equilibrium before drinking. At this new equilibrium, samples were taken for TRLFS measurements, pH and TIC were re-determined and speciation calculations were performed according to the new analytical data. The emission peaks of U(VI) in CLASSIC waters showed no differences in comparison with STILL waters (Table 4-4). However, the bi-exponential function which was used to fit the fluorescence decays of these waters indicated, beside the main ternary complex $Ca_2UO_2(CO_3)_{3(aq)}$, a minor species with shorter fluorescence lifetime. Moreover, analysis of the individual emission peaks showed a spectral shift with increasing the delay time which demonstrated a second species (see Fig.4-6). Referring to Table (4-4), these short lifetimes can be compared to the lifetime of $UO_2CO_{3(aq)}$ as a dominant species in a model solution. Nevertheless, due to the composition of the medium and the possible existence of other U(VI) complexes the observed lifetime of $UO_2CO_{3(aq)}$ could be reduced from 465 μs to that values obtained for CLASSIC waters. This assumption was supported by speciation modeling of the water as shown in figure (4-3). The binding form $UO_2CO_{3(aq)}$ is found as a minor species in the total distribution of U(VI), whereas $Ca_2UO_2(CO_3)_{3(aq)}$ is dominating.

In both types of mineral waters $Ca_2UO_2(CO_3)_{3(aq)}$ was confirmed as a major binding form, while $UO_2CO_{3(aq)}$ was assigned as a secondary species in CLASSIC water due to the decrease in pH values which changed the U(VI) distribution. Beside the pH value the Ca^{2+} concentration and TIC played the major role as master parameters that control uranium(VI) species distribution. This has already been elucidated by the speciation modeling of the mineral waters.

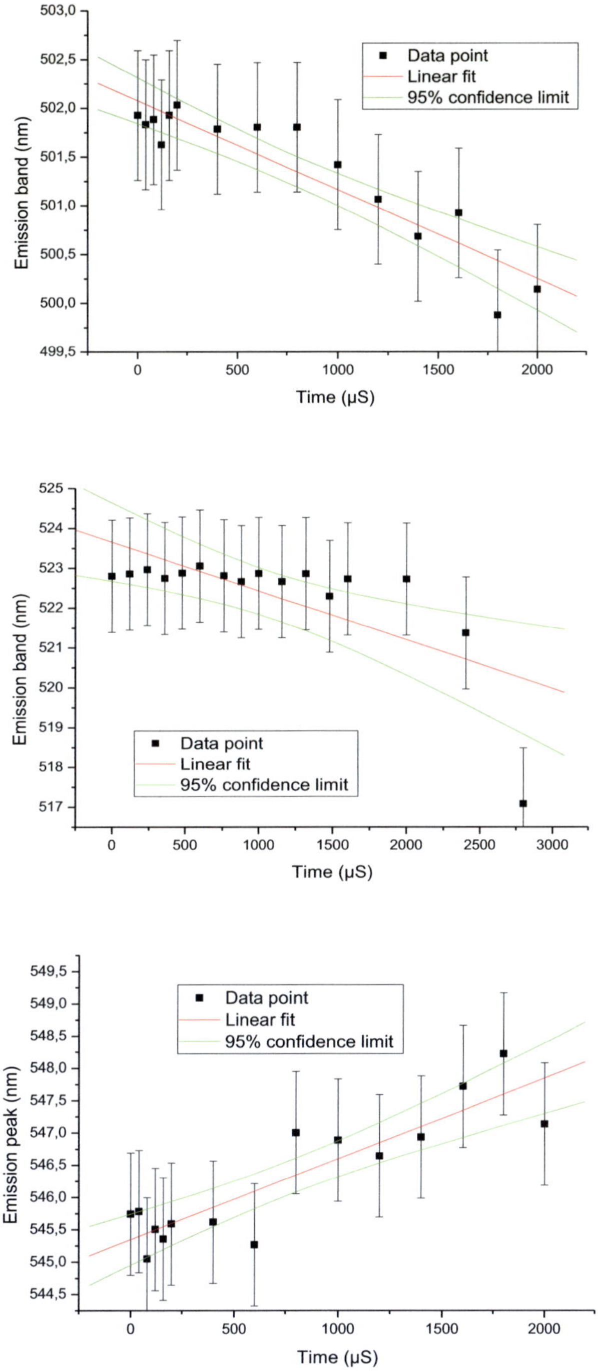

Figure 4-6. Analysis of individual emission peaks as a function of delay time in Teinacher as an example of CLASSIC water

4.1.2 U(VI) speciation in groundwater – Pofadder case study

Here we investigated uranium speciation in some naturally contaminated groundwater with high uranium content from Pofadder, South Africa. Uranium levels in most waters in that area exceed the WHO guideline value and were reported to be in the range > 10 – 300 µg/L, and in few cases more than 300 µg/L (Winde 2011). Speciation data of uranium could give valuable information for epidemiological studies linked to such elevated U-levels in drinking water.

4.1.2.1 Sampling and water parameters

Five groundwater samples were taken from five boreholes in Pofadder rural arid area in the Northern Cape Province of South Africa near Namibia. Main water parameters (cations and anions) and U concentrations were determined by elemental analysis techniques (ICP-MS and IC). Total organic carbon (TOC) was quantified by a total carbon analyzer. Water samples were prepared in the form of ice block, as in the case of mineral waters, and were measured using cryo-TRLFS. Details about laser fluorescence measurements and the analysis techniques are in the section "Experimental Details".

4.1.2.2 Speciation modeling of Pofadder groundwater

As in the case of mineral waters, the U speciation in Pofadder groundwater samples were also calculated by the geochemical code EQ3/6 (Wolery 1992) using the stability constants which are listed in Table (4-2).

Table 4-5. Uranium binding forms in the groundwater from Pofadder arid area calculated by the geochemical speciation code EQ3/6 using NEA-TDB

Sample ID	**Uranium binding form (%)**		
	$Ca_2UO_2(CO_3)_{3(aq)}$	$CaUO_2(CO_3)_3^{2-}$	$MgUO_2(CO_3)_3^{2-}$
Spieelpan (S-pan)	98.5	-	-
Poortijie (P)	97.4	-	1.2
Grappies (GRP)	98.8	-	-
Calcrete wind pump (CC)	99.1	-	-
Potchefstroom (pot)	95.3	-	3.2

Table 4-6. Concentrations of some water quality parameters (inorganic elements) and total organic compounds – Pofadder, South Africa

Sample ID	pH	U(VI) (µg/L)	Na^+ (mg/L)	Mg^{2+} (mg/L)	Ca^{2+} (mg/L)	F^- (mg/L)	Cl^- (mg/L)	SO_4^{2-} (mg/L)	HCO_3^- (mg/L)	TOC (mg/L)
Spieelpan (S-pan)	7.93	478	301	88.6	219	1.97	622	291	300.9	1.02
Poortijie (P)	7.91	13.7	122	20.1	66.6	3.42	122	54.9	317.2	0.88
Grappies (GRP)	7.58	135	1080	49.5	333	2.93	1580	731	154	0.77
Calcrete wind pump (CC)	7.33	432	4900	799	2010	3.36	10400	4150	248.6	23.9
Potchefstroom (pot)	8.18	4.04	24.9	44.3	61.2	< 0.1	32	93.5	282	2.12

Table 4-7. Luminescence characteristics of U(VI) in groundwater samples under investigation measured at 153 K, $\lambda_{ex.} = 266$ nm

Sample code	**Spectral maxima (nm)**	**τ(µS)**	**Decay function**	**Main U species**
Spieelpan (S-pan)	481.1, 501.0, 522.0, 544.6, 570.0	938±6	Mono-exponential	$Ca_2UO_2(CO_3)_{3(aq)}$
Poortijie (P)	482.3, 502.0, 522.6, 545.3, 570.6	1033±36	Mono-exponential	$Ca_2UO_2(CO_3)_{3(aq)}$
Grappies (GRP)	481.4, 501.2, 521.9, 543.7, 570.5	974±11	Mono-exponential	$Ca_2UO_2(CO_3)_{3(aq)}$
Calcrete wind pump (CC)	482.1, 501.5, 522.4, 545.3	957±9	Mono-exponential	$Ca_2UO_2(CO_3)_{3(aq)}$
Potchefstroom (pot)	481.9, 501.2, 522.6, 545.2	991±30	Mono-exponential	$Ca_2UO_2(CO_3)_{3(aq)}$
Synthetic $Ca_2UO_2(CO_3)_{3(aq)}$	481.0, 501.4, 522.4, 545.0	1009±11	Mono-exponential	$Ca_2UO_2(CO_3)_{3(aq)}$

4.1.2.3 Luminescence data of uranium in Pofadder groundwater

The uranium binding forms of the investigated waters was experimentally checked by monitoring the U(VI) luminescence with the cryo-TRLFS at 153 K. All uranium species in the samples were non-specifically excited by 266 nm delivering energy of 5 mJ in a 10-ns pulse and with a repetition rate of 20 Hz (Details on the measurement and TRLFS setup are found in the section "Experimental Details"). U(VI) luminescence data (Table 4-7) exhibited almost similar characteristics as in mineral waters (STILL type). The spectra were best fitted with a mono-exponential decay function (Annex III) indicating the sole presence of the highly soluble species $Ca_2UO_2(CO_3)_{3(aq)}$. In other words, elevation in uranium concentration does not influence its binding forms at the pH range of drinking waters. This was confirmed by a synthetic solution of $Ca_2UO_2(CO_3)_3$ consists of 1 mM Ca^{2+}, 5 mM HCO_3^-, and 1 μM U(VI) at pH 7.2 (Table 4-7). However, such high U-level has the advantage of faster TRLFS measurement with a better bands resolution in comparison to mineral waters with U-levels < 2 μg/L. In both mineral and ground waters, the Ca^{2+} concentrations control uranium speciation and increase its mobility by forming the soluble complex $Ca_2UO_2(CO_3)_{3(aq)}$.

4.1.3 U(VI) speciation in acid drainage waters

Two acidic effluent samples were investigated for uranium binding forms in wastewaters that are produced by human activities. Such information could be useful in the framework of waste repositories and/or in the framework of restoration of contaminated sites that were affected by mining or industrial activities.

4.1.3.1 Samples locations and analytical data

The first sample was taken from untreated effluent of an old superphosphate fertilizer factory (Kynoch, South Africa). The water is characterized by a low pH of 1.4 and relatively high content of phosphate, sulfate, and natural organic matter (Table 4-8). As uranium is naturally associated with phosphate minerals and ores, it is found at relatively high concentration in the untreated effluent of the factory. The second sample, which is used for comparison, was a mine water from the former uranium mine of Königstein in Saxony, Germany. The sample is also acidic (pH 2.6) with very high uranium concentration due to the in-situ leaching by sulfuric acid in the past. Analytical data (Table

4-8) was determined by both ICP-MS and IC, details about both techniques are given in the section "Experimental Details".

Table 4-8. Analytical data of the two acid drainage waters under investigation

Component	Concentrations in mM	
	Wastewater – Kynoch	Mine water – Königstein
Na^+	9.6	5.8
K^+	1.8	0.27
Ca^{2+}	20.9	5.9
Mg^{2+}	3.8	0.66
U	0.0004	0.0334
SO_4^{2-}	20.7	11
PO_4^{3-}	32.8	0.014
Cl^-	3.3	4.2
TOC (mg/L)	15.9	2.9
pH	1.4	2.6

4.1.3.2 Thermodynamic modeling of acid drainage waters

The chemical behavior of such acid waters is influenced by pH value and the presence of ligands that tend to bind U(VI) in acidic media such as sulfate, phosphate, and chloride. The speciation calculation for both waters demonstrated $UO_2SO_{4(aq)}$ as a major binding form whereas the free uranyl ion (UO_2^{2+}) also existed in a notable percentage (Table 4-9).

Table 4-9. Calculated U(VI) binding forms in the investigated acid waters using the NEA-TDB and geochemical speciation code EQ3/6

Sample ID	Uranium(VI) binding form (%)		
	$UO_2SO_{4(aq)}$	UO_2^{2+}	$UO_2(SO_4)_2^{2-}$
Wastewater – Kynoch	47.5	47.5	3.2
Mine water – Königstein	69.4	25	5.5

4.1.3.3 Luminescence data of U(VI) in acid drainage waters

TRLFS measurements, at both room and cold temperature, were performed to determine U(VI) species in the acid waters under investigation. Effluent water from Kynoch showed no luminescence at room temperature. This was attributed not only to the low U content (4×10^{-7} M) but mainly to the presence of luminescence quenchers such as the high content of the natural organic matter (TOC: 16 mg/L) as well as the relatively high chloride concentration. Contrary to this, Königstein water showed a clear and a well-resolved spectrum and the observed uranium species produced emission bands at 476.1, 492.9, 514.8, 538.4, 564.4 nm which indicated the presence of uranium sulfate complexes.

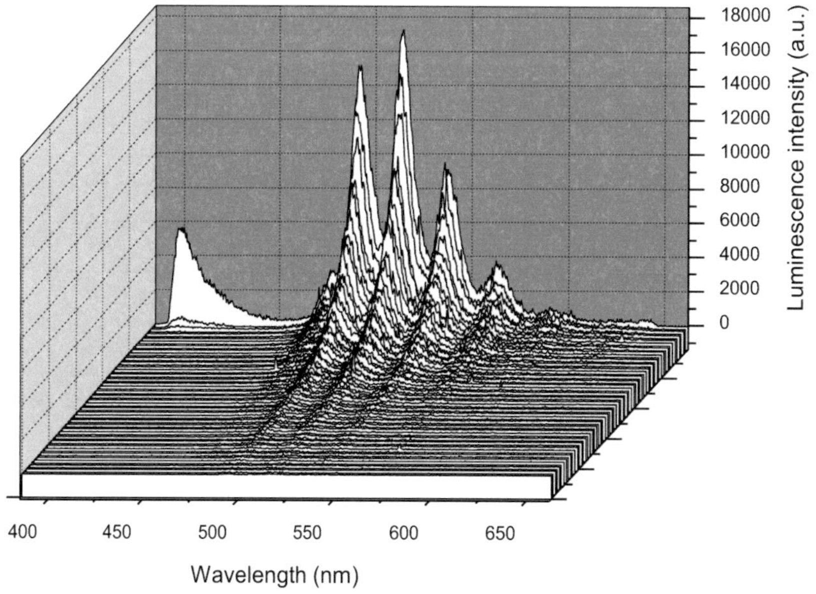

Figure 4-7. TRLFS spectrum at 25°C of acid mine water from Königstein, Germany

Applying the cryo-TRLFS for both samples, interesting features of the resulted spectra were observed. The spectral information of both waters indicated no U(VI) – sulfate species, but mainly the luminescence characteristics of uranyl – chloride system which was reported by Osman and Geipel (2012). Spectroscopic results either at room temperature or at low temperature are not consistent with thermodynamic prediction. For instance, luminescence data of U(VI) in Königstein water at room temperature indicated only uranium sulfate complexes while the predicted free uranyl ion was undetectable from the spectral information. This could be due to the presence of chloride as a fluorescence quencher.

Table 4-10. Spectroscopic data at both room and cold temperature of the two acid drainage waters and a model solution of U(VI) – chloride system

Sample ID	**Emission bands (nm)**	**Lifetime (µs)**
Wastewater – Kynoch (at room temperature)	No uranium luminescence	-
Mine water – Königstein (at room temperature)	476.1, 492.9, 514.8, 538.4, 564.4	0.158±0.015, 0.713±0.011
Wastewater – Kynoch (at low temperature)	491.9, 513.1, 536.4, 561.4	428.4±7.5
Mine water – Königstein (at low temperature)	491.0, 512.2, 535.1, 560.5, 588.3	422.4±12
U(VI) – chloride system (at low temperature)	491.7, 513.0, 536.6, 562.7, 590.0	451.9±5.9

Furthermore, luminescence data for both samples at low temperature (153 K) reflected only U(VI) – chloride system although the thermodynamic prediction indicated no uranyl chloride complexes. This can be explained by quenching mechanism of chloride ions of fluorescence at room temperature which includes electron transfer from the quencher (Cl^-) to the fluorophore (excited uranyl $^*UO_2^{2+}$) with formation of a non-fluorescent species (UO_2Cl^+). As a consequence, the intensity and lifetime of the excited uranyl ($^*UO_2^{2+}$) is shortened or completely vanished. However, lowering the temperature reduces the collisional process and the complex ($^*UO_2Cl^+$) can stay longer in the excited state resulting in longer lifetime and higher luminescence yield even much greater than the U(VI) complexes with SO_4^{2-}, PO_4^{3-}, and OH^-.

4.1.4 Concluding remarks

Application of laser fluorescence spectroscopy coupled with a cold temperature system (cryo-TRLFS) provides a remarkably good capability for direct determination of uranium speciation in environmentally relevant waters at extremely low levels such as in mineral waters or at exceptionally high concentrations such as in mining or ground waters, with the potential to be developed further. The chemical species $Ca_2UO_2(CO_3)_{3(aq)}$, which is known as a less toxic species (Carriere et al. 2004), was found to be the main binding form that dominated uranium distribution in both mineral and ground waters by means of both spectroscopy and thermodynamic modeling. So, depending on the proven quality of the used thermodynamic data, modeling could be a good tool to understand and to monitor the behavior of uranium in natural aquatic systems. Uranium luminescence data of acid drainage waters reflect the importance of chloride ions even in the presence of other stronger ligands such as sulfate and phosphate.

Data about uranium binding forms in such environmentally relevant waters could be used for more insight research on uranium chemo-toxicity in drinking water since there is no available information on the long-term exposure of *low uranium concentrations* via ingestion on humans. The only published data was the nephrotoxicity in rats after a short-term and a high dose uranium exposure (Carriere et al. 2005). Therefore, it could be a baseline for epidemiological studies in the case of drinking waters of naturally high uranium-levels such as in Pofadder arid area. Moreover, such knowledge could also give a good perspective on uranium speciation in human body fluids after incorporation. Last but not least, speciation data about drainage waters which is produced by human activities might be valuable for remediation strategies on contaminated sites (old uranium mines) and/or designing concept for waste repositories.

4.2 Interaction of uranium(VI) with bioligands present in human biological fluids: The case study of urea and uric acid

4.2.1 Relevance of the study

With the present increase of uranium accumulation in the food chain and eventually in the bio-system including humans, knowledge about uranium binding forms with bioligands present in biological fluids at natural relevant concentrations is needed and will be useful to understand its toxicity, behavior and distribution in such cellular level as well as for evaluating toxicity after uranium exposure. There is a lack of pertinent data to support thermodynamic calculations of U(VI) in human biological fluids, especially with relevant biological ligands. It would be more meaningful when considering these organic substances even with their weak complexes in the equation model for a reliable and exact prediction. For this purpose we investigate the complexation of uranium(VI) with bioligands (urea and uric acid), which commonly exist in considerably high concentrations in biological fluids such as urine, sweat, and blood plasma. Although other minor bioligands (citrate, oxalate, glycine, etc.) with stronger complexing properties might dominate the binding of U(VI), still it would be better for a complete picture to see how urea and uric acid behave with U(VI) at such natural relevant concentrations and in a such complicated biological matrix. Furthermore, determination of the associate complex formation constant of such systems will be of great interest since it provides a good understanding and/or modeling the transport behavior of uranium(VI) in such biological aqueous systems. Furthermore, calculating uranium(VI) speciation in human urine as an example of a biofluid gives a clear picture on the effect of such relevant biological ligands.

4.2.2 Bioligands under investigation

Urea and uric acid are known as the most and major metabolites in biological systems. Both have been quantified in different body fluids by many authors (Liu et al. 2012; Huang et al. 2002; Song and Hou, 2002; Marsh et al. 1965). Urea (CH_4N_2O) is the main end-product of metabolism of nitrogenous compounds (proteins and amino acids) in animals. It is produced by the liver via urea cycle, and most of it is secreted via urine sooner after entering the blood system. In addition to that it has an important role in the metabolism of toxic ammonium in the gastrointestinal tract after circulating in the blood system (Williams and Lansford, 1967). Uric acid ($C_5H_4N_4O_3$) is a heterocyclic organic molecule that is created when the body breaks down purine nucleotides i.e. it is the final and major oxidation (breakdown) product of catabolism of the purine nucleotides, adenosine and guanosine (Huang et al. 2002). The chemical structure of both molecules is shown in figure (4-8).

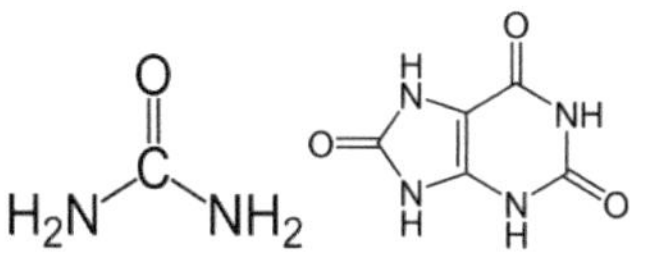

Figure 4-8. Structural formulae of the studied ligands

Due to their biological importance, both compounds were studied thoroughly, especially from the biochemical viewpoint. As a bioligand, uric acid interacts with bio-metal ions such as Mg, Ca, Fe, Co, Cu, and Zn to form soluble and/or solid complexes (Moawad 2002; Davies et al. 1986). Its salts and complexes have also been obtained with some other cations; Mn(II), Fe(II), Co(II), Ni(II), Cu(II), Zn(II), Al(III), Fe(III) (Moawad 2002), and with Cr(III) as well (Moawad 2002; Ramana et al. 1992). Some thermodynamic data of some urea complexes with Mn(II), Ni(II) and Zn(II) in organic medium were determined by Ozutsumi et al. (1995) using titration calorimetry. Recent data from Heller et al. (2009) showed complexation of urea in aqueous solution with europium(III) as a rare earth metal and with curium(III) as a radioactive element using TRLFS. Up to date no data is available about interaction and behavior of urea and uric acid with uranium as a hazardous element that accumulates in the biosphere at present and eventually in the living organisms. This has been achieved by the means of laser spectroscopy in line with thermodynamic calculations using the computer codes HYDRA and MEDUSA (Puigdomenech 2004). All experiments were performed by a ns-pulsed laser using 266 nm wavelength with a pulse energy of 0.3 mJ for exciting the uranyl complexes. The TRLFS spectra were recorded in the range from 371.39 to 674.28 nm averaging three time-resolved spectra with 50 laser shots for each single spectrum. The gate time was set to 2 µs, 20 µs, and the delay varied from 150 – 10000 ns, 50 – 40000 ns for uric acid and urea, respectively. Technical details on the ns-TRLFS system are given in the section “Experimental Details”.

4.2.3 Interaction of uranium(VI) with uric acid

Figure (4-9) shows the luminescence spectrum of a 5×10^{-5} M uranium(VI) at pH 3.0 as a function of uric acid concentrations. Binding to urate caused a decrease in the uranyl luminescence intensity which indicates a static fluorescence quenching and formation of a non-fluorescent complex as well. The mono-exponential decay indicated the presence of the free uranyl ion with a lifetime of 1.40±0.08 µs. A decrease in the lifetimes was also observed with increasing uric acid concentrations indicating an additional dynamic quench effect (Table 4-11).

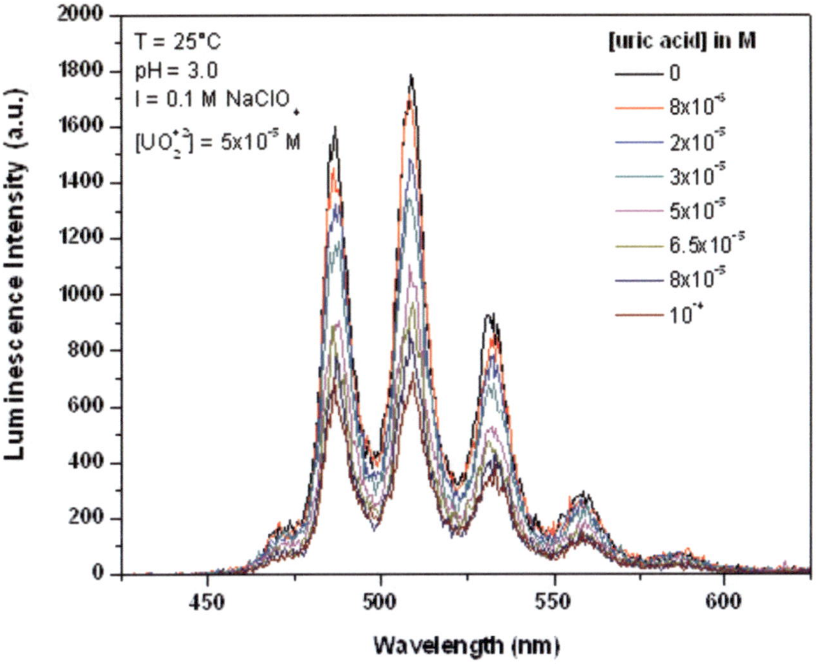

Figure 4-9. Luminescence spectra of U(VI) at pH 3.0 at various uric acid concentrations (no peak shift)

Since the formed complex was non-fluorescent; the formation constant was calculated by a fluorescence quenching experiments using TRLFS. In this method, according to Stern and Volmer (1919), the term (I_0/I)-1 is plotted against the concentration of the quencher (uric acid); where (I_0/I) is the ratio of the fluorescence intensities before and after addition of uric acid and the slope is used to derive the stability constant log β = 4.36±0.01 (Fig.4-10). Furthermore, the straight line in figure (4-10) indicates a 1:1 stoichiometry. Applying slope analysis method using mass action law; stoichiometry of the reaction can also be estimated and the stability constant can be validated (Fig.4-11). The concentrations of the free uranyl ion were determined from the measured luminescence intensities. These concentrations and their corresponding luminescence intensities were used to calculate the corresponding concentrations of the uranyl-urate complex and the non-complexed ligand.

Table 4-11. Luminescence lifetimes for uranyl – uric acid system at pH 3

[uric acid] in M	**Lifetime (τ) in ns**	**[uric acid] in M**	**Lifetime (τ) in ns**
0	1396±8	5×10^{-5}	1248±9
8×10^{-6}	1375±6	6.5×10^{-5}	1231±10
2×10^{-5}	1341±6	8×10^{-5}	1117±17
3×10^{-5}	1264±7	10^{-4}	913±19

Uric acid $C_5H_4N_4O_3$ is known as a diprotic acid (can be refer to here as H_2Ur). Taking into consideration the protonation of uric acid at pH 3, it tends to form acid urate (i.e. the single charged hydrogen $C_5H_4N_4O_3^-$ or HUr^-) and dissociates according to (Eq.4-3):

$$H_2Ur \leftrightarrow H^+ + HUr^- \quad (4-3)$$

Then the first dissociation constant (Ka_1) of uric acid can be defined as in (Eq.4-4):

$$Ka_1 = \frac{[H^+][HUr^-]}{[H_2Ur]} \qquad (4-4)$$

Depending on the above fact, uranyl ion (UO_2^{+2}) reacts with the acid urate (HUr^-) as follows in (Eq.4-5):

$$UO_2^{2+} + n\,HUr^- \leftrightarrow \{(UO_2)(HUr)_n\}^{(2-n)+} \qquad (4-5)$$

$$\beta = \frac{[\{(UO_2)(HUr)_n\}^{(2-n)+}]}{[UO_2^{2+}][HUr^-]^n} \qquad (4-6)$$

Equation (4-6) can be converted into a linear form by taking logarithm values for both sides (Eq.4-7):

$$\log\frac{[\{(UO_2)(HUr)^n\}^{(2-n)+}]}{[UO_2^{2+}]} = n\cdot\log[HUr^-] + \log\beta \qquad (4-7)$$

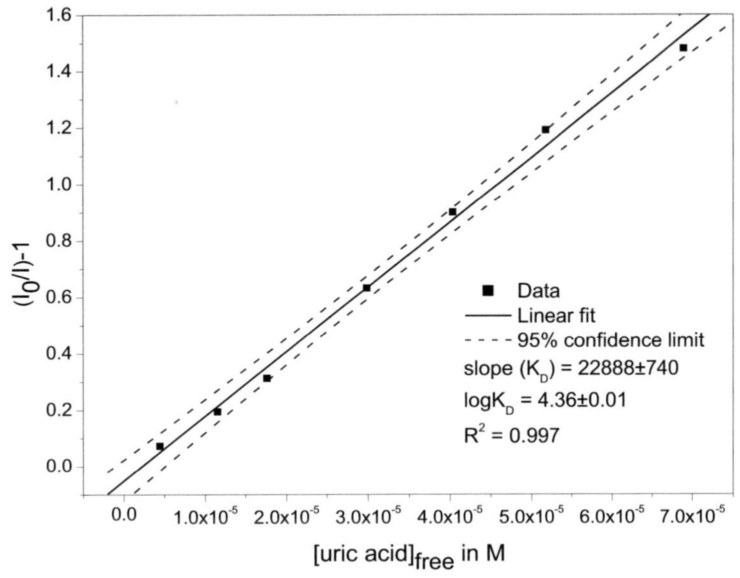

Figure 4-10. Fluorescence quenching experiment, static quenching of uranyl ion by uric acid at 25°C

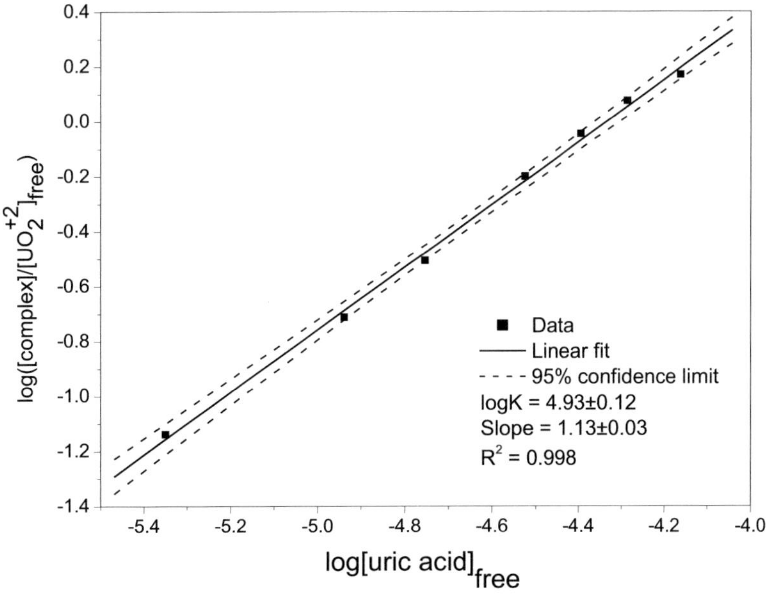

Figure 4-11. Slope analysis of the complex reaction of uranium(VI) - uric acid system

Using equation (4-7), slope analysis was performed to estimate the stoichiometry of equation (4-5) (see Fig.4-11). A slope of 1.13±0.03 indicates a 1:1 complexation, the corresponding formation constant was obtained from the intercept; log β = 4.93±0.12. The factor analysis program SPECFIT was also applied for the determination of log β (details about SPECFIT is found in the section "Experimental Details"). Concentrations of UO_2^{2+} and H_2Ur, pH, and dissociation constant (p*Ka*) of H_2Ur were used in SPECFIT calculations as input parameters. We used the p*Ka* that was determined by Biltz and Herrmann (1921). The calculated log β via SPECFIT was 4.72±0.15 and this is compatible with other values as summarized in Table (4-12).

Table 4-12. Summary of methods used to determine log β for uranium(VI) – urate system using TRLFS data

Calculation method	**Stern-Volmer**	**Slope analysis**	**SPECFIT**	**Average**
Log β_{110}	4.36±0.01	4.93±0.12	4.72±0.15	4.67±0.29

We see that urate presents itself as a complexing agent for U(VI) as well as for other transition metals. This property plays an important role in biological systems such as human blood; a good example is the inhibition of ascorbate oxidation and lipid peroxidation by urate – iron binding (Davies et al. 1986). Uric acid in blood acts as a scavenger for all transition metal ions including uranium, *if presents*, which enter biosystem and has a negative effect by oxidizing ascorbic acid and generating free radicals. In other words, uric acid has an antioxidant property via complexing with such metal ions. The U(VI) – urate formation constant (log β = 4.67) can be compared to that of 1:1 urate – Fe(II) complex (log β = 4.28) and both are significantly less than the 1:1 (log β = 5.38) and 1:2 (log β = 5.65) of urate – Fe(III) complexes as reported by Davies et al. (1986). Based on the newest data (log β_{110} = 4.67±0.29) of U(VI) – uric acid system, we calculated uranium(VI) species distribution in aqueous solution of a pH range similar to that of bio-fluids (3 to 7), as indicated in figure (4-12).

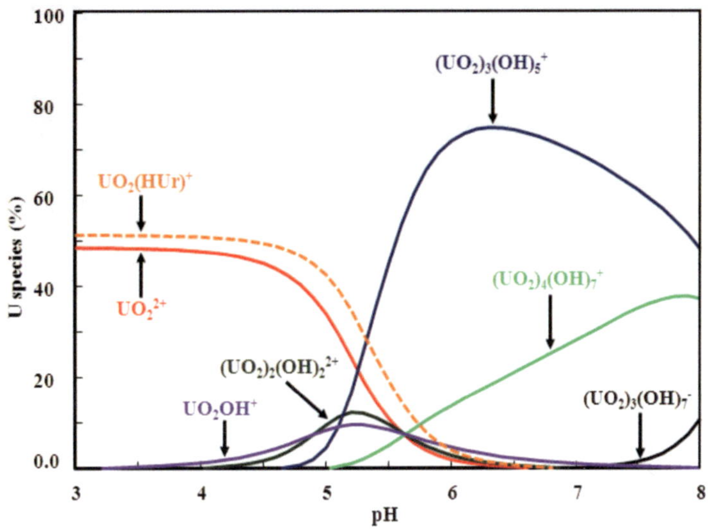

Figure 4-12. MEDUSA – Species distribution of U(VI) as a function of pH in a system containing 50 μM UO_2^{2+}, 50 μM uric acid, at $I = 0.1$ M ($NaClO_4$) and 25°C

4.2.4 Interaction of uranium(VI) with urea

Here two sets of ligand concentrations were used. The first set ranged from 10^{-5} to 10^{-4} M; the second set represents relatively high concentrations of urea 0.01 to 0.3 M which more similar to the range of urea in human urine (Heller 2011) and human sweat (Huang et al. 2002). Each set of urea concentrations behaves differently when interacts with a 5×10^{-5} M uranyl solution in a medium of pH 4 and $I = 0.1$ M ($NaClO_4$) at room temperature.

4.2.4.1 Urea set 10^{-5} to 10^{-4} M

Figure (4-13(a)) shows the luminescence spectrum of a 5×10^{-5} M uranium(VI) at pH 4.0 as a function of urea concentrations. Similar to the case of uric acid, static fluorescence quenching and formation of a non-fluorescent complex was observed. The bi-exponential decay of TRLFS spectra indicated the presence of a mixture of the free uranyl ion with a lifetime of 1.40±0.02 μs and a uranyl hydroxide species mostly $(UO_2)_2(OH)_2^{2+}$ with a lifetime of 9.3±0.04 μs. Eliet et al. (1995) reported this species $(UO_2)_2(OH)_2^{2+}$ with almost the same lifetime of 9.5±0.3 μs. The fluorescence quenching experiment was also used to calculate the formation constant (log β) for the uranium(VI) – urea system (Fig.4-14(a)), and log β was found to be 3.63±0.05. To estimate the stoichiometry; slope analysis was done (Fig.4-15(a)). The slope of 1.07 gave an evidence of a 1:1 complexation in this system with log β = 3.93±0.36. The calculated log β via SPECFIT was 3.80±0.31 which is consistent with other values as shown in Table (4-13).

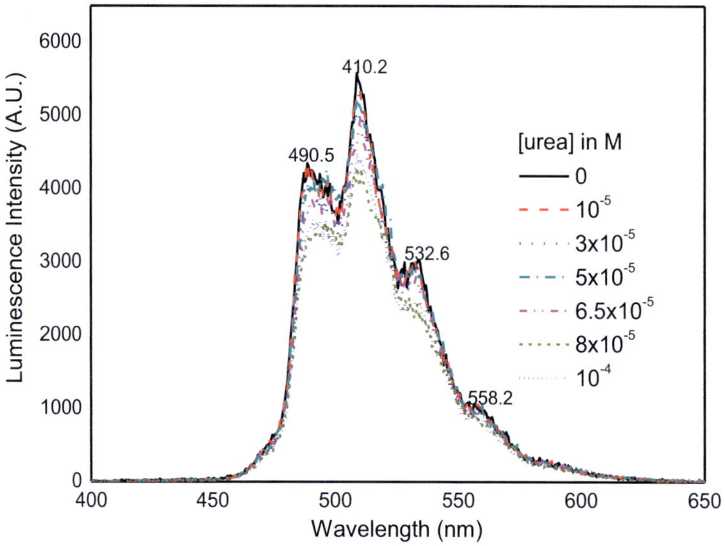

(a)

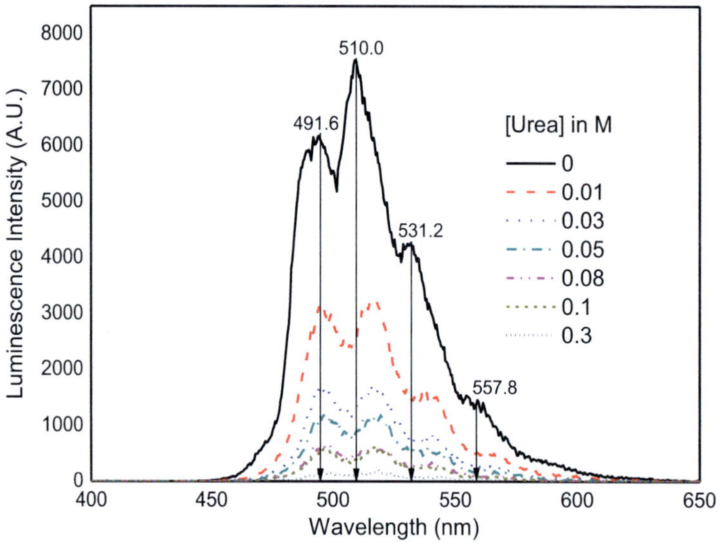

(b)

Figure 4-13. Static quench effect of urea on the aqueous free UO_2^{2+} luminescence, **(a)** set1, 10^{-5}-10^{-4} M urea **(b)** set2, 0.01-0.3 M urea

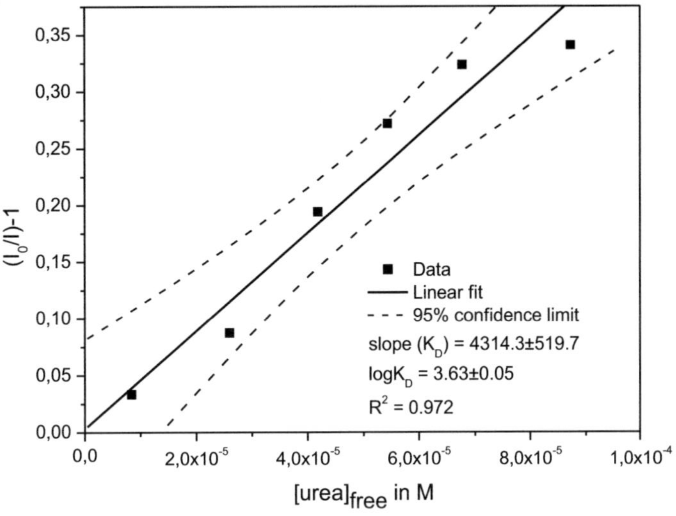

Figure 4-14(a). Fluorescence quenching experiment, static quenching of uranyl ion by urea at 25°C

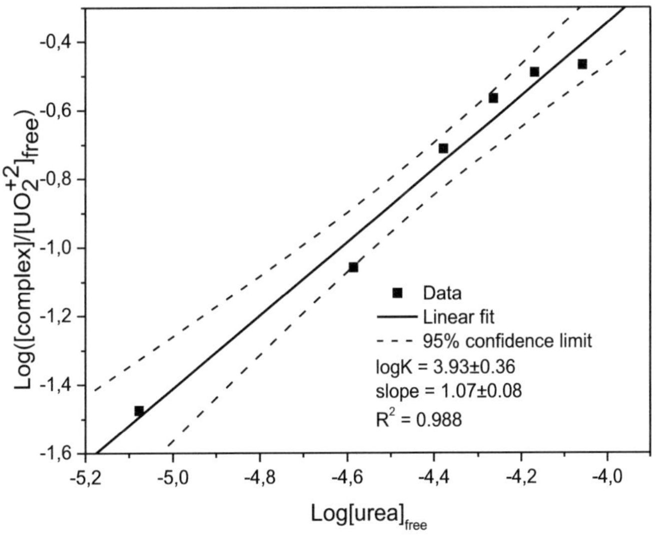

Figure 4-15(a). Slope analysis of the complex reaction of uranium(VI) - urea system

4.2.4.2 Urea set 0.01 to 0.3 M

Referring to (Fig.4-13(b)) we observed that the emission maxima were more red-shifted in comparison to urea set 10^{-5} – 10^{-4} M (Fig.4-13(a)). This might be due to the contribution of the relatively high concentrated urea which enhanced the ionic strength of the solution and consequently increased the activity of the uranyl hydroxyl species. For this urea set (0.01 – 0.3 M); the stability constant (log β) was also obtained using different methods as for the first set (see Fig.4-14(b) and Fig.4-15(b)). Both urea sets gave significantly different formation constants with uranyl ion (see Table 4-13) at the same reaction conditions as stated previously.

Table 4-13. Summary of methods used to determine log β for uranium(VI) – urea system using TRLFS data

Calculation method	**Stern-Volmer**	**Slope analysis**	**SPECFIT**	**Average**
Log β_{110}: urea set1 (10^{-5} – 10^{-4} M)	3.63±0.05	3.93±0.36	3.80±0.31	3.79±0.15
Log β_{110}: urea set2 (0.01– 0.3 M)	2.28±0.02	2.14±0.10	1.93±0.15	2.12±0.18

This difference in log β for both urea complexes can be attributed to the higher urea concentrations in the second set (0.01 – 0.3 M) which has a combined effect. The first effect is the strong quenching of urea on free aqueous uranyl ion. The second effect is the increasing of ionic strength which enhanced uranyl to form hydroxyl species more than to bind to urea. Further explanation can be clarified when we look back to some properties of urea especially its structure. Urea is a very weak base with a relatively high permanent dipole moment and hence electrolytically conducting in its aqueous solution. This electrolytic behavior can be well understood by a zwitterion structure. It is reported in the literature (Harris and Robson, 1948; Venkatesan and Suryanarayana, 1956; Bergmann and Weizmann, 1938) that urea mostly exists as a zwitterion NH_2^+:C(NH_2)O^- and the following resonating hybrid structure were suggested:

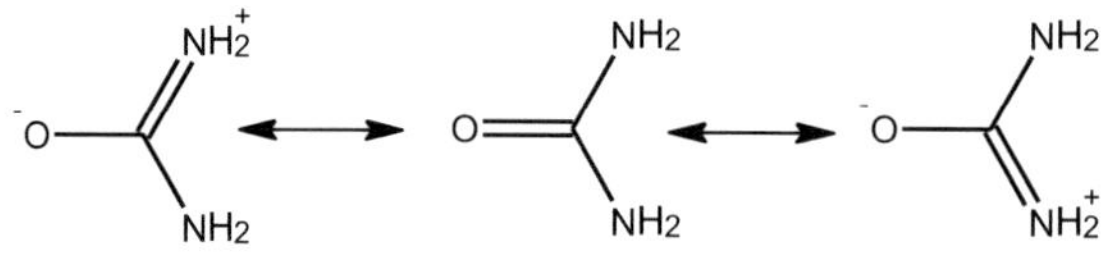

Figure 4-16. Suggested resonating hybrid structure of urea

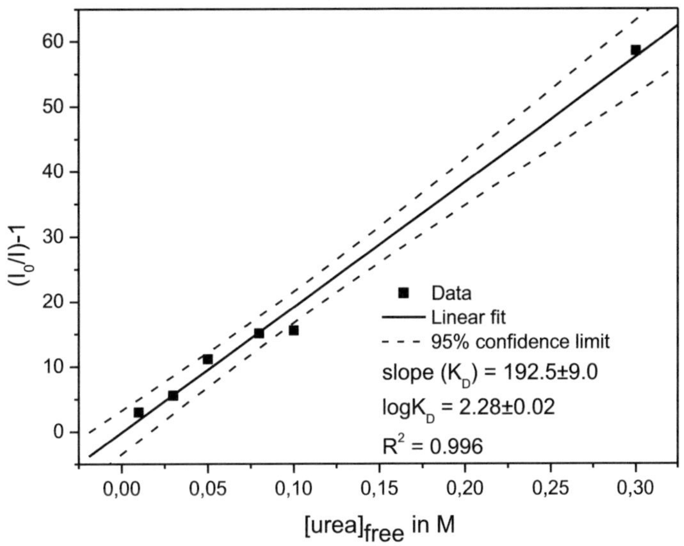

Figure 4-14(b). Fluorescence quenching experiment, static quenching of uranyl ion by urea at 25°C

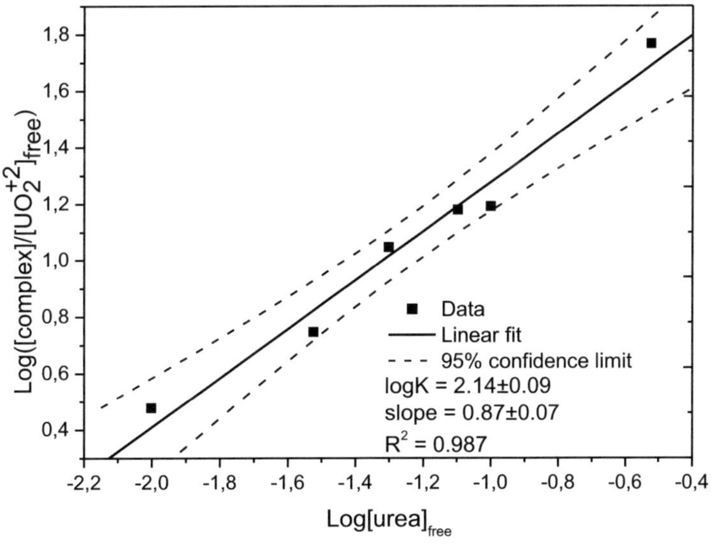

Figure 4-15(b). Slope analysis of the complex reaction of uranium(VI) - urea system

Accordingly, if we have more urea molecules in solution, the UO_2^{2+} will pose more repulsion from $-NH_2^+$ and this repulsion is low at lower urea concentrations. Furthermore, intermolecular interactions (van der Waals and hydrogen bonding) getting stronger with increasing the number of urea molecules and hence the solution deviates from ideality (i.e. in concentrated liquids ions are not randomly distributed as we see in the case of ideal solutions). Therefore, usage of a molar concentration of urea set (0.01 – 0.3 M) is not practical for calculating log β. For more accurate calculations activities rather than concentrations are needed.

Generally, urea complexation with uranyl can be written as (Eq.4-8):

$$x\ UO_2^{2+} + y\ {}^{+}H_4N_2CO^{-} \leftrightarrow \{(UO_2)_x(H_4N_2CO)_y\}^{2x} \quad (4-8)$$

We expect that urea is coordinated to U(VI) via the lone pair of electron at the oxygen atom of the carbonyl group (>C=O:→U) as approved by van Staveren et al. (1987). As for U(VI) – uric acid system; uranium species distribution in aqueous solutions was also calculated (see Fig.4-17) depending upon our newest data of U(VI) – urea system for both relatively low (set 1) and relatively high (set 2) urea concentrations.

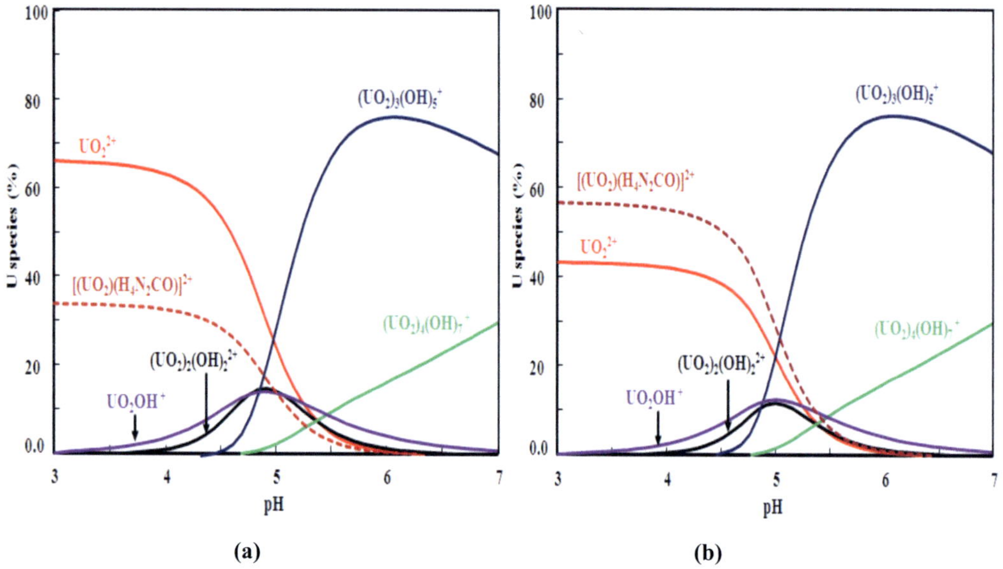

Figure 4-17. MEDUSA – U(VI) species distribution as a function of pH in a system containing 50 µM UO_2^{2+}, 0.1 mM urea **(a)**, 10 mM urea, **(b)** at I = 0.1 M ($NaClO_4$) and 25°C

4.2.5 U(VI) speciation calculations in human urine

Investigating the behavior of U(VI) in the biosystems (e.g. biofluids, cell fluids) will be one important application of the studied complexes. In the following example U(VI) speciation was calculated based on inorganic composition of human urine (Table 4-14) used by Sutton and Burastero (2003) as well as organic composition (Table 4-15).

Table 4-14. Inorganic constituents of human urine used by Sutton and Burastero (2003)

Element	**µM**	**Element**	**µM**
Na^+	174075.16	F^-	84.21
K^+	64850.81	Cl^-	146666.67
Mg^{2+}	4681.11	CO_3^{2-}	10^4
Ca^{2+}	5285.46	SO_4^{2-}	99375
U(VI)	1	PO_4^{3-}	57674.25

pH range 4.2 – 8.0, $I = 0.7$ M

U(VI) speciation calculation in urine was done considering only inorganic constituents used by Sutton and Burastero (2003). Referring to figure (4-18) we see that phosphate complexes, namely, UO_2HPO_4 is the prevailing U(VI) species till pH 6 then $Ca_2UO_2(CO_3)_{3(aq)}$ starts to become the prominent species up to pH around 9. Above pH 9 tricarbonato complex $UO_2(CO_3)_3^{4-}$ dominates the U(VI) speciation. This seems a bit contradicted to what Sutton and Burastero (2004) calculated. They showed that U(VI) bound to phosphate as a $UO_2(HPO_4)_2^{2-}$ throughout the pH range under investigation and never indicated $Ca_2UO_2(CO_3)_{3(aq)}$ which mostly dominates U(VI) speciation in solutions of pH (6 – 9) with Ca^{2+} concentrations greater than 1 mM. In the second case we consider the most possible organic constituents in urine that formed complexes with U(VI) and their thermodynamic stability constants were already reported. These reported thermodynamic data and our newly generated data were included in the equation model beside inorganic data and then used for new calculations for U(VI) speciation in urine. As depicted in figure (4-19), U(VI) speciation was tremendously changed due to the presence of the indicated bioligands. The glycine complex $UO_2(Hgly)_2^{2+}$ dominates the U(VI) distribution in acidic urine, at pH > 4.5 citrate begins to show higher affinity to bind U(VI) than glycine till pH 6.5. Then in neutral and slightly alkaline urine samples still the carbonato complexes ($Ca_2UO_2(CO_3)_{3(aq)}$ and $UO_2(CO_3)_3^{4-}$) are dominating the distribution. Although other bioligands are present at relatively high concentrations such as urea, they play an insignificant role in U(VI) speciation in urine, due to the weak complexation with uranium (e.g. $\log \beta = 2.12$ for U(VI) – urea). Nevertheless, in pure aqueous solutions urea and uric acid complexes dominate the speciation specifically $UO_2(HUr)^+$ (Fig.4-20(a)). In original urine samples and in the absence of glycine and citrate still both urea and uric acid have no significant role in U(VI) speciation, only $UO_2(HUr)^+$ can be seen in a higher percentage at a pH < 3.5 (Fig.4-20(b)).

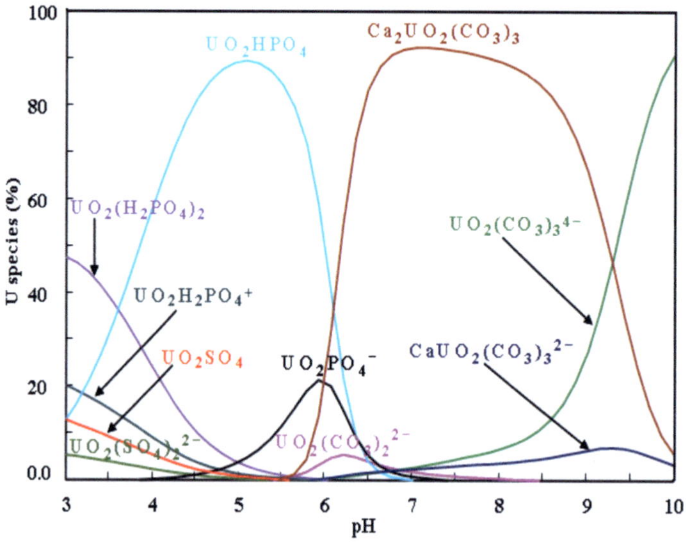

Figure 4-18. MEDUSA – U(VI) speciation as a function of pH in human urine according to inorganic constituents used by Sutton and Burastero (2004)

But both bioligands could play a role in U(VI) distribution in other biofluids such as human sweat. Although the formation constants for both U(VI) – uric acid and U(VI) – urea systems were determined at pH 3 and 4 respectively; but they are still valid for speciation calculation in the biological pH range of human urine (4 – 8), since both ligands do not undergo deprotonation in the indicated range of pH.

Table 4-15. Some important main organic constituents of human urine (Heller 2011)

Organic compound	mM	Organic compound	mM
Citric acid	0.5 – 4.0	Uric acid	1.1 – 5.4
Oxalic acid	0.06 – 1.0	Urea	100 – 500
Lactic acid	0.5 – 4.0	Glycine	1.0 – 2.0

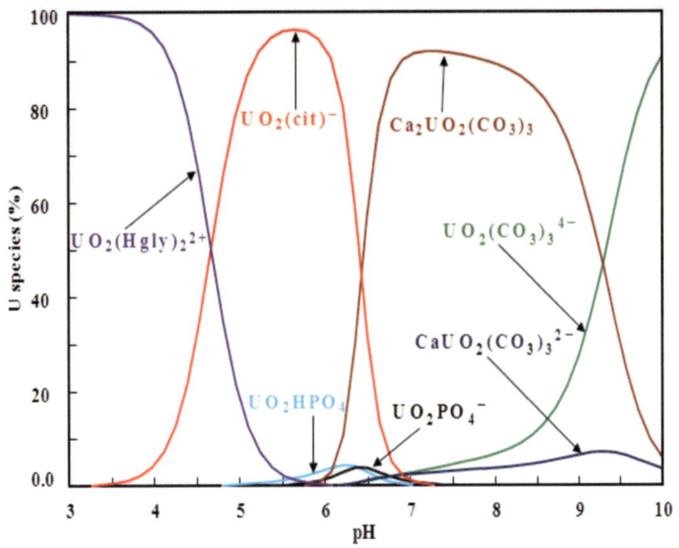

Figure 4-19. MEDUSA – U(VI) speciation as a function of pH in human urine based on the main organic and inorganic constituents

4.2.6 Concluding remarks

In this study the formation constants of uranium(VI) with both uric acid and urea in aqueous solution were reported for the first time. Both complexes are fairly weak, since they provide only N-H and kentones functionalities which are known to form weak complexes with hard Lewis acids like uranium. The formation constant of uranyl – uric acid system was higher than that of uranyl – urea system. This might be due to the mode of complexation; as we find uric acid forms chelate with uranyl ion, urea represents a monodentate ligand. The major aim of this study was to provide information about U(VI) complexation with some basic organic constituents (uric acid, urea) of human biological fluids at natural relevant concentrations. Following uranium metabolic pathways after incorporation/exposure through its speciation in the body fluids by spectroscopic means is so difficult, since it is not possible to see all uranium complexes or species. This is because uranium complexation with inorganic ligands dominates the spectra and gives no chance to non-luminescent complexes formed by uranium with other bio-organic ligands such as urea, uric acid, creatinine, and citrate which represent considerable part of human bio-fluids. Based on our newest data compilation and other reported data; U(VI) binding forms were calculated in human urine. The speciation calculation showed the influence of organics with carboxylic functionalities, mostly glycine and citrate, at pH < 6.5, but they had no effect at higher pH range (7 – 9), while other organics such as uric acid and urea are insignificant on the speciation.

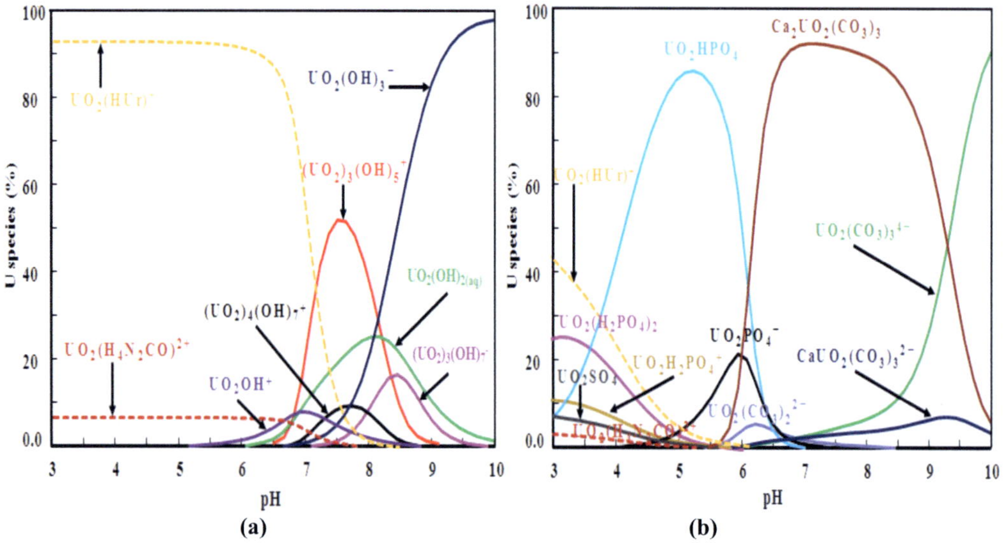

Figure 4-20. MEDUSA – Influence of uric acid and urea on U(VI) distribution as a function of pH in **(a)** pure aqueous solution **(b)** human urine based only on the main inorganic constituents

It is certain that bioligands with carboxylic functionalties have greater binding potential and can play a basic role of U(VI) distribution in biosystem. Some have already been studied and published, and were included in our speciation calculations to see to what extend can they compete with urea and uric acid for U(VI) distribution in human urine. Obviously, our findings indicate that both investigated bioligands unlikely play a major role in U(VI) distribution in human urine comparing to other bioligands with carboxylic groups or chelating properties (refer to Fig.4-19 and Fig.4-20). Nevertheless, taking into account organic constituents, regardless of their weak complexation, in the equation model beside inorganic constituents is of great importance and contribute in a better understanding and/or accurate calculations and simulations of the intricate uranium speciation in human biological fluids. Uranium(VI) speciation in biological fluids by means of both spectroscopy and thermodynamic modeling are detailed in the next section.

4.3 Uranium(VI) binding forms in some human body fluids: spectroscopy and modeling investigation

4.3.1 Significance and aim of the study

Due to the increase of uranium content in the bio-sphere at present, it is expected to be progressively accumulated in the living organisms including humans. As previously mentioned, uranium may pose chemical and/or radiological risk to human health in the case of internal contamination (inhalation, ingestion, wound). The behavior of uranium (intake, uptake, distribution, excretion) inside the body which is termed as its biokinetics (Roth et al. 2001) is governed by its binding forms or the so-called chemical speciation. Knowledge about uranium binding forms (speciation) in natural body fluids will be useful not only to understand its toxicity and distribution in the cellular level, but also is of primary importance for designing decorporation therapy in the case of accidental overexposure. In this context, uranium speciation in the body fluids was somewhat highlighted during the last two decades. For instance, Scapolan et al. (1997; 1998a) characterized some uranium complexes in biological medium and their application to blood serum by means of capillary electrophoresis (CE) and time-resolved laser-induced fluorescence spectroscopy (TRLFS). Sutton and Burastero (2004) performed thermodynamic modeling of uranium distribution in several simulated human biological fluids. The reported simulated fluids contain significantly less constituents than the natural biological fluids as they did not include organic constituents; they only considered the interaction with inorganic constituents in their model. However, there is insufficiency of relevant data to support such models, particularly the complexation of uranium with bioorganic ligands that commonly exist at relatively high concentrations in the body fluids. Therefore, considering interaction of uranium(VI) with the relevant biological ligands is of great importance for an accurate model and/or a better understanding of uranium biokinetics. Previous work about uranium(VI) binding forms in the body fluids was only limited to thermodynamic speciation calculations, and there is no spectroscopic verification of the obtained data. Moreover, samples were also limited to simulated solutions which contain significantly less constituents than natural aqueous system or body fluids. It could be possible to learn more from nature by the use of spectroscopic identification of unknown species in real samples obtained from a human body. The purpose of this investigation is to enhance the previous knowledge about uranium speciation in body fluids by combining both theoretical modeling and spectroscopic methods in both artificial and original samples (saliva, sweat, urine). The challenge is to provide experimental data in such complicated matrices in line with theoretical thermodynamic calculations.

4.3.2 Specimens' collection/preparation

Body fluids under investigation (saliva, sweat, urine) were collected as spontaneous samples from three healthy volunteers each. Saliva and urine specimens were taken after breakfast, while sweat was collected during a heat stress in a dry sauna facility. In the latter case the volunteers took a shower before getting inside the sauna. Fluorescence spectroscopic measurements were performed on the same day after spiking all samples with uranium. Based only on the inorganic ions concentrations of the collected natural specimens and literature findings (Marques et al. 2011; Heller et al. 2011; Guyton and Hall, 2000; Iyengar et al. 1978), artificial body fluids were prepared. Analytical grade reagents and CO_2-free Milli-Q water were used for the preparation of all solutions. Details on preparation and chemicals are given in the section "Experimental Details".

4.3.3 Analytical data and uranium spiking experiments

All natural body fluids and artificial solutions were analyzed for inorganic composition (cations and anions) and the total organic carbon (TOC) without delay using elemental analysis techniques (ICP-MS and IC). Figure (4-21 (a, b, c)) presents the inorganic composition of natural specimens as well as artificial solutions (details are found in Annex (V)). The pH values were measured immediately after specimens' collection.

A uranium stock solution of 5×10^{-3} M $UO_2(ClO_4)_2$ in 0.02 M $HClO_4$, whose exact concentration was checked by ICP-MS, was used. Small aliquots (3 mL) from each body fluid was spiked with uranium solution to get final concentrations of 1 µM U(VI) in both saliva and sweat and 50 µM U(VI) in urine. No change in the pH values was observed due to the buffering capacity of the body fluids (HCO_3^-/CO_3^{2-} buffer, $H_2PO_4^-/HPO_4^{2-}$ buffer). Artificial body fluids were also spiked with the same concentrations, but different pH values were set for each fluid within the range of their biological pH. In other words, pH-dependent experiments on U(VI) speciation were performed only for artificial fluids. On the other hand,

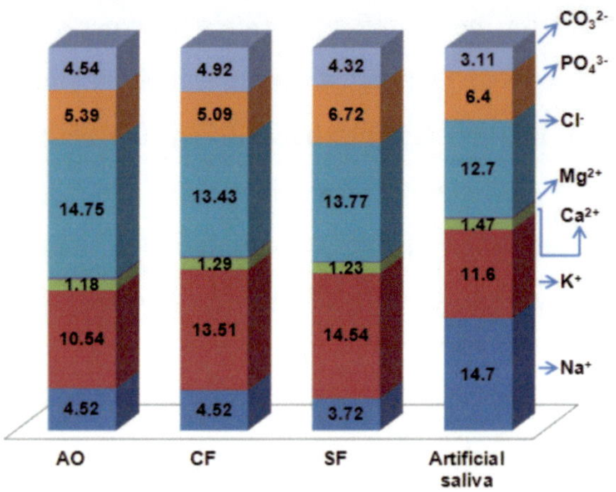

Figure 4-21(a). Inorganic composition (in mM) of both natural human saliva (AO, CF, SF) and artificial solution

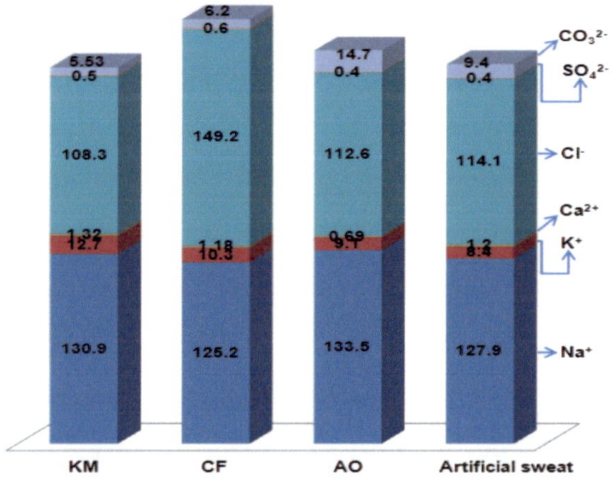

Figure 4-21(b). Inorganic composition (in mM) of both natural human sweat (KM, CF, AO) and artificial solution

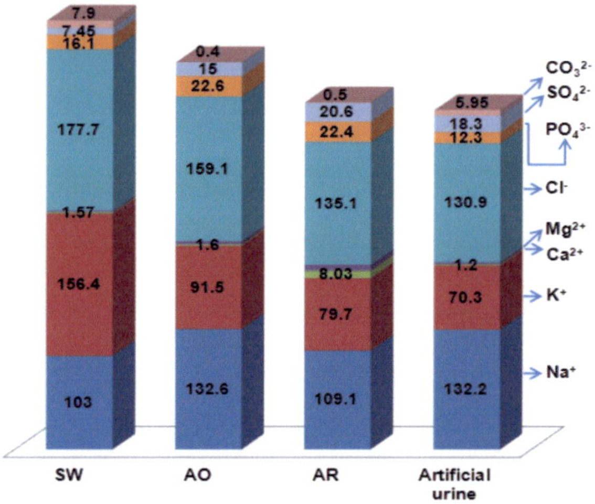

Figure 4-21(c). Inorganic composition (in mM) of both natural human urine (SW, AO, AR) and artificial solution

speciation measurements for the natural body fluids were performed at their original pH values. All spiked-samples were then shock frozen using liquid nitrogen and then stored in a deep freezer (-20°C) and eventually extracted from the cuvette as an ice block and measured by cryogenic time-resolved laser-induced fluorescence spectroscopy (cryo-TRLFS) on the same day after spiking all samples with uranium. Details on sample preparation for laser measurements, cryo-TRLFS, and the analysis techniques are given in the section "Experimental Details".

4.3.4 U(VI) speciation in body fluids – modeling

Distribution of uranium(VI) binding forms in the body fluids under investigation was theoretically calculated by the computer codes HYDRA and MEDUSA (Puigdomenech 2004) using the most updated thermodynamic stability constants of aqueous uranyl complexes which were compiled by the Nuclear Energy Agency – Thermochemical Database (NEA-TDB) (Guillaumont et al. 2003). Other non-compiled data, but are of significance in the aqueous chemistry of uranium, are the earth alkaline uranyl carbonate were also used. These include the aqueous $Ca_2UO_2(CO_3)_3$, $CaUO_2(CO_3)_3^{2-}$ (Bernhard et al. 2001), and $MgUO_2(CO_3)_3^{2-}$ (Dong and Brooks, 2006). These complexes were not used by Sutton and Burastero (2004) in their calculations of U(VI) speciation in simulated human biological fluids. Details on the values and sources are shown in Table (4-2). Besides that aqueous complexation of U(VI) with main organic constituents in the body fluids with their respective thermodynamic stability constants were also included in the model. Speciation of U(VI) was calculated in both artificial solutions and the original body fluids considering; 1) only inorganic constituents, and 2) both inorganic and main organic constituents. Parameters which have been used in the calculations were pH, ionic strength, ionic concentrations, and organic ligands concentrations. Uranium(VI) concentration of 1 µM was considered for both saliva and sweat and 50 µM was considered for urine.

4.3.4.1 Speciation modeling of U(VI) in saliva

U(VI) distribution in artificial and original saliva fluids was calculated considering only inorganic composition at first. Here U(VI) is typically distributed as phosphate species ($UO_2HPO_{4(aq)}$, $UO_2PO_4^-$, $UO_2H_2PO_4^+$) below pH 6.3 (Fig.4-22(a)). The presence of Ca^{2+} makes the aqueous species $Ca_2UO_2(CO_3)_3$ prominent at pH > 6.3 (Fig.4-22(a)). Accordingly, we expect that the fluorescence spectroscopic measurements of these fluids will show the typical spectra of $Ca_2UO_2(CO_3)_{3(aq)}$ at biological pH of 6 to 7. However, the fluorescence data did not indicate the aqueous $Ca_2UO_2(CO_3)_3$ but showed fluorescence bands that are similar to complexes of earth alkaline uranyl phosphate system which was reported by Geipel (2005b). This was further discussed in our spectroscopic findings.

$$Ca^{2+} + 2UO_2^{2+} + 2PO_4^{3-} = Ca(UO_2)_2(PO_4)_{2(aq)},\ \mathrm{Log}\,\beta_{122} = 35.30\ (I = 0.1\mathrm{M}) \qquad (4-9)$$

$$Ca^{2+} + UO_2^{2+} + PO_4^{3-} = CaUO_2PO_4^{+},\ \mathrm{Log}\,\beta_{111} = 16.62\ (I = 0.1\mathrm{M}) \qquad (4-10)$$

$$Mg^{2+} + 2UO_2^{2+} + 2PO_4^{3-} = Mg(UO_2)_2(PO_4)_{2(aq)},\ \mathrm{Log}\,\beta_{122} = 33.44\ (I = 0.1\mathrm{M}) \qquad (4-11)$$

Taking the aforementioned aqueous complexes in the calculations, U(VI) distribution was strikingly changed as the positively charged species $CaUO_2PO_4^{+}$ dominated at the biological pH (6 – 7) of human saliva (Fig.4-22(b)).

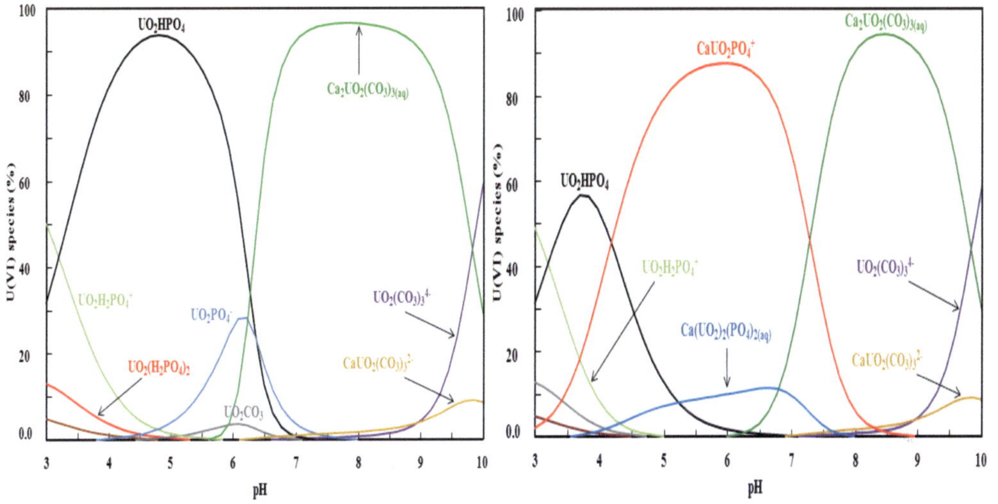

Figure 4-22. U(VI) species distribution as a function of pH in artificial saliva without (**a**), and with (**b**) considering data from Geipel (2005b)

On the other side, human saliva is rich with organic biomolecules and proteins of which α-amylase is the most important as it is representing 40 to 50% of salivary protein (Oppenheim et al. 2007; Noble 2000). Moreover, it contributes in the initial digestion of dietary starch within the oral cavity (Hoebler et al. 1998). Thus, α-amylase was considered in the speciation calculation taking the mean concentration value of 2.64 mg/mL which was assayed by Mandel et al. (2010). Thermodynamic formation constants of α-amylase-U(VI) system were taken from Barkleit and Hennig (2013). As presented in equations (4-12) and (4-13) U(VI) can bind to one or two carboxylic groups in α-amylase, respectively.

$$UO_2^{2+} + [(COO)\text{-}Amy] = UO_2[(COO)\text{-}Amy]^{2+}, \; \text{Log}\,\beta_{110} = 5.8 \qquad (4\text{-}12)$$

$$UO_2^{2+} + [(COO)_2\text{-}Amy] = UO_2[(COO)_2\text{-}Amy]^{2+}, \; \text{Log}\,\beta_{120} = 10.6 \qquad (4-13)$$

All natural saliva specimens revealed almost the same distribution for U(VI) as in the artificial saliva when only inorganic constituents were taken into account (Fig.4-22(b) and Fig.4-23(a)). However, including α-amylase as the most abundant protein in natural human saliva changed the speciation greatly. In the biological pH range of human saliva (6 to 7), the complex of which uranium bound to two carboxyl groups in α-amylase $UO_2[(COO)_2\text{-}Amy]^{2+}$ dominated the speciation until pH around 6.6 then the positively charged complex $CaUO_2PO_4^+$ was the major binding form till pH 7.0 (Fig.4-23(b)). Nevertheless, contribution from other minor U(VI) species can be seen in the indicated pH range as well.

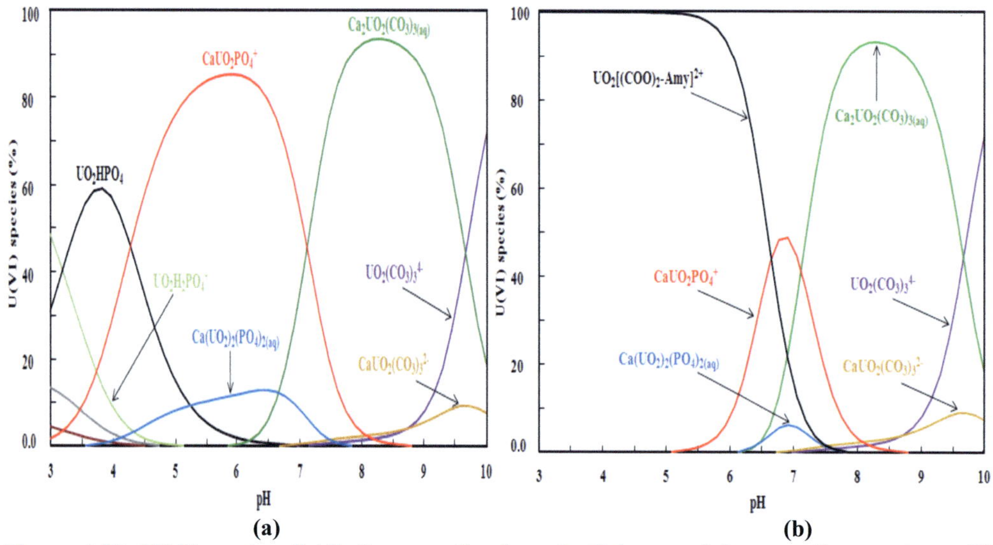

Figure 4-23. U(VI) species distribution as a function of pH in natural human saliva specimen (SF) considering (**a**) only inorganic composition, (**b**) both inorganic and organic composition

4.3.4.2 Speciation modeling of U(VI) in sweat

No differences were observed in U(VI) speciation in both artificial and natural human sweat (Fig.4-24) in view of inorganic composition only. However, influence of pH was clear since human sweat has a wide biological pH range (4.0 – 6.8 for eccrine sweat and 6.0 – 7.5 for apocrine sweat) (Draelos 2010; Robinson and Robinson, 1954). The binding form UO_2CO_3 dominated in slightly acidic sweat (pH around 5 to 6), while $Ca_2UO_2(CO_3)_{3(aq)}$ is the major binding form in neutral and alkaline pH range.

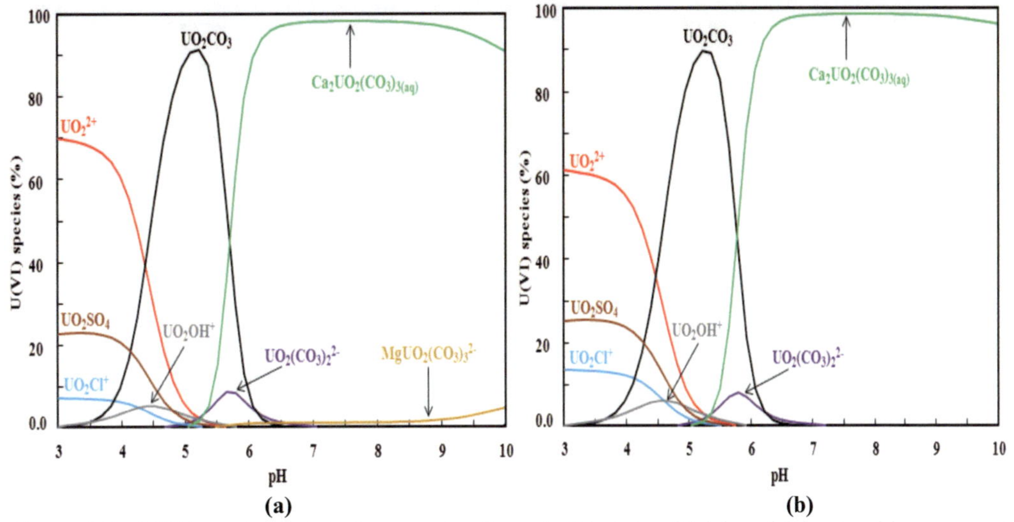

Figure 4-24. U(VI) species distribution as a function of pH in (**a**) artificial, and (**b**) natural human sweat specimen (CF) considering only inorganic composition

In natural sweat specimens the speciation was recalculated taking into account the major organic composition (lactate, urea, uric acid) beside inorganic composition. Typical concentrations of urea (2.20 mM) and uric acid (25 μM) in human sweat (Huang et al. 2002) and their complex formation constants with U(VI) as reported by Osman et al. (2013b) were used. Lactate levels have a wide variation in human sweat depending on the region of the body (Patterson et al. 2000) and are influenced by age and gender as well (Meyer et al. 2007). Nonetheless, we took the average sweat lactate concentration of 22.1 mM for a heat trial and 43.7 mM for endurance physical training (Derbyshire et al. 2012). Thermodynamic stability constants for lactate-U(VI) system were taken from Schmeide et al. (2012). It is obvious that U(VI) distribution was affected by the presence of organics (mainly lactate) in acidic sweat until pH 5.5, then $Ca_2UO_2(CO_3)_{3(aq)}$ dominated the speciation beyond pH 6 (Fig.4-25).

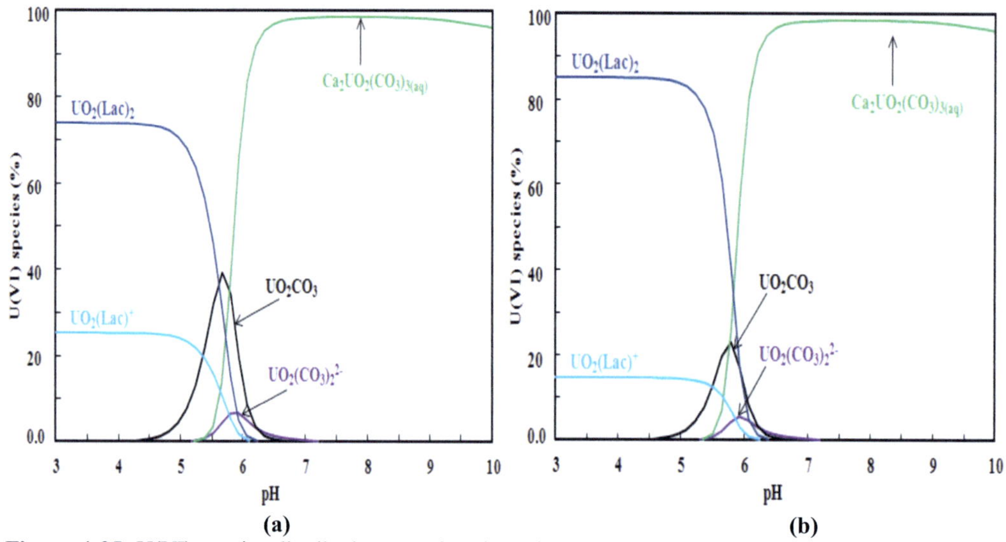

Figure 4-25. U(VI) species distribution as a function of pH in natural human sweat specimen (CF) with average sweat lactate concentration (**a**) 22.1 mM, and (**b**) 43.7 mM

4.3.4.3 Speciation modeling of U(VI) in urine

In artificial urine the phosphate complexes (mainly UO_2HPO_4) prevailed U(VI) species until pH 6, then the aqueous $Ca_2UO_2(CO_3)_3$ dominated at neutral and alkaline pH range (Fig.4-26(a)). Fluorescence spectroscopy did not show the characteristics of those species, but mainly the characteristics of the system $Ca^{2+}/UO_2^{2+}/PO_4^{3-}$ which was reported by Geipel (2005b). As in the case of saliva, thermodynamic data for the system $Ca^{2+}/UO_2^{2+}/PO_4^{3-}$ was included in the equation model. Accordingly, U(VI) distribution in the pH range 4.5 to 7 was governed by the species $Ca(UO_2)_2(PO_4)_{2(aq)}$ as a major binding form and $CaUO_2PO_4^+$ as a minor species, and then $Ca_2UO_2(CO_3)_{3(aq)}$ predominated at pH > 7 (Fig.4-26(b)). In view of inorganic composition only, the artificial urine has a similar speciation diagram (Fig.4-26(b)) as the natural urine specimen (SW) (Fig.4-27(a)) but different from the other two natural specimens (AO) and (AR) (Fig.4-27(c, e)). The dissimilarity was attributed to the low carbonate content and relatively high phosphate levels in both (AO) and (AR) specimens. As a consequence, the complex $Ca(UO_2)_2(PO_4)_{2(aq)}$ predominated even beyond pH 7.5 (Fig.4-27(c, e)). This is in practice unrealistic since the indicated complex has extremely low solubility and precipitates at pH > 7.5. Natural urine specimens are varied not only in their ionic concentrations but also in their organic composition which depends mainly on the nutritional habits.

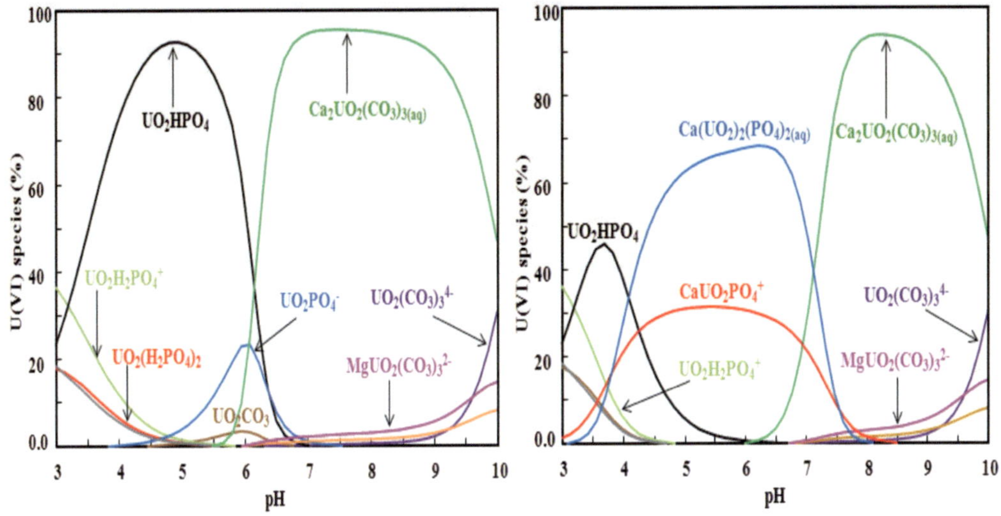

(a) (b)

Figure 4-26. U(VI) species distribution as a function of pH in artificial urine without (**a**), and with (**b**) considering data from Geipel (2005b)

Table (4-16) shows the main organic composition of human urine and their natural relevant concentrations (by range) (Heller 2011) as well as their concentrations that were used in the speciation calculations. Considering aqueous complexation of U(VI) with the main organic constituents of human urine with their respective thermodynamic stability constants (Table 4-17) changed the speciation tremendously (Fig.4-27).

Table 4-16. Main organic constituents of human urine and their natural relevant concentrations

Bioorganic ligand	**Concentrations (mM)**	
	Human urine	**Values used for speciation modeling**
Citric acid (Cit^{3-})	0.5 – 4.0	2.0
Oxalic acid (Ox^{2-})	0.06 – 1.0	0.5
Lactic acid (Lac^-)	0.5 – 4.0	2.0
Uric acid (H_2Ur)	1.1 – 5.4	3.0
Urea (H_4N_2CO)	100 – 500	200
Glycine (HGly)	1.0 – 2.0	1.0

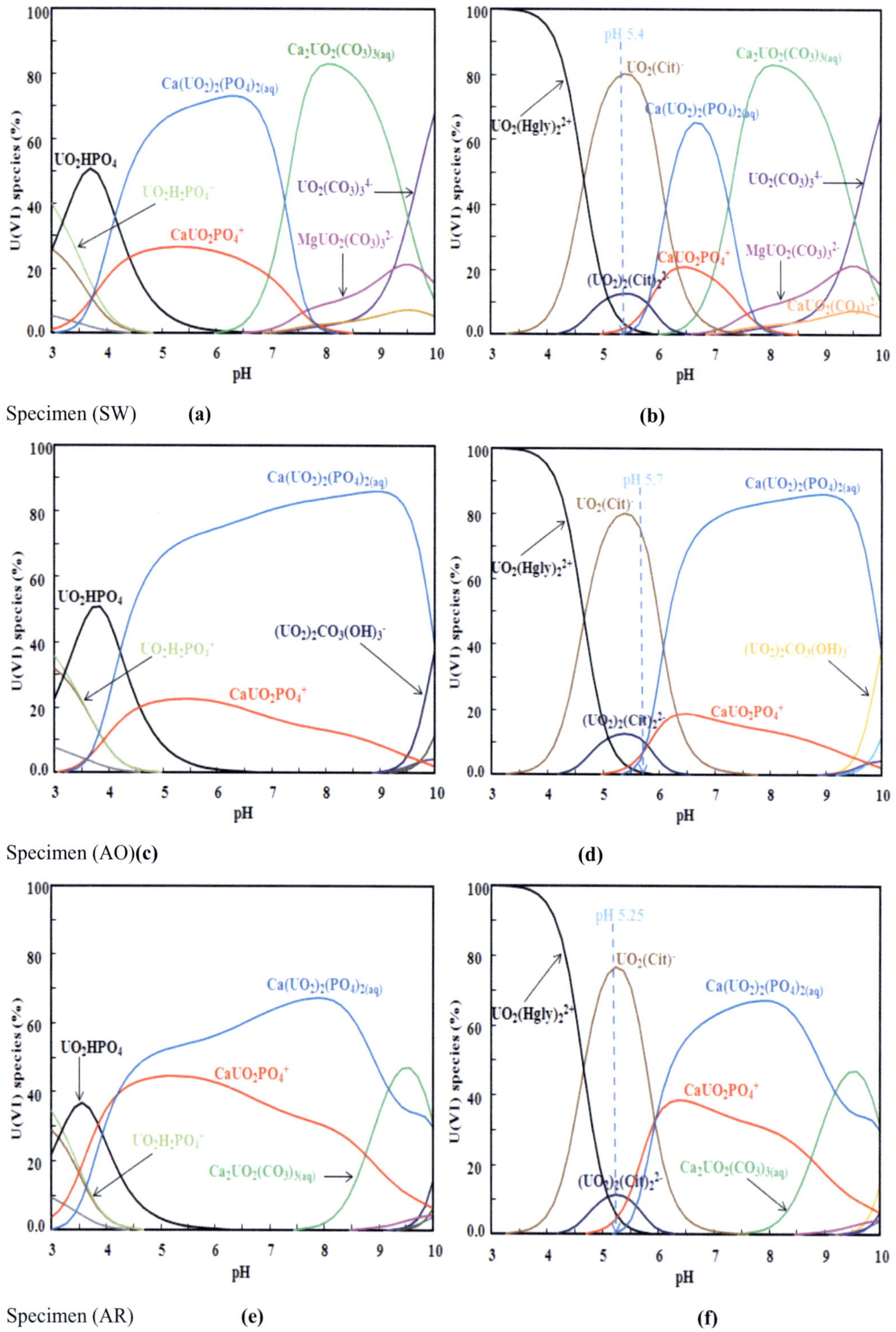

Figure 4-27. U(VI) species distribution as a function of pH in natural human urine specimens considering only inorganic composition (**a, c, e**), and both inorganic and organic composition (**b, d, f**)

For instance glycine complex $UO_2(HGly)_2^{2+}$ prevailed in acidic urine (pH < 4.5) whereas citrate represented an important complexant for U(VI) at pH range 4.5 to 6. Above pH 6 inorganic ions governed U(VI) speciation depending mainly on their concentrations. In a carbonate rich urine (e.g. specimen SW) the aqueous $Ca_2UO_2(CO_3)_3$ prevailed at pH ≥ 7.5 while in carbonate poor urine (specimens AO and AR) the $Ca(UO_2)_2(PO_4)_{2(aq)}$ predominated even above pH 7.5 depending on Ca^{2+} and UO_2^{2+} concentrations. Despite their high concentrations in human urine, both urea and uric acid are insignificant in U(VI) distribution comparing to other bioligands (glycine and citrate). This is attributed to their weak complexation with U(VI).

Table 4-17. Complexation of U(VI) with some bioligands commonly exist in human urine

Reaction	**Log $\beta_{(I=0.1M, T=25°C)}$**	**Reference**
$UO_2^{2+} + Cit^{3-} = UO_2Cit^-$	7.67±0.12	Guenther et al. 2011
$2UO_2^{2+} + 2Cit^{3-} = (UO_2)_2(Cit)_2^{2-}$	18.85±0.42	Guenther et al. 2011
$UO_2^{2+} + ox^{2-} = UO_2ox$	5.92±0.03	Guenther et al. 2011
$UO_2^{2+} + 2ox^{2-} = UO_2(ox)_2^{2-}$	10.30±0.38	Guenther et al. 2011
$UO_2^{2+} + HUr^- = UO_2HUr^+$	4.67±0.29	Osman et al. 2013b
$UO_2^{2+} + H_4N_2CO = UO_2(H_4N_2CO)^{2+}$	2.12±0.18	Osman et al. 2013b
$UO_2^{2+} + H^+ + Gly^- = UO_2(HGly)^{2+}$	10.84±0.1	Guenther et al. 2007
$UO_2^{2+} + 2H^+ + 2Gly^- = UO_2(HGly)_2^{2+}$	21.19±0.13	Guenther et al. 2007

The previous speciation models provide just an idea on how U(VI) is distributed in such biological matrices. Although the aqueous complexes of U(VI) with the indicated bioligands were not compiled by the NEA-TDB, they provide information about their possible effect. Accordingly, spectroscopic data is strongly needed to support the aforementioned calculation data.

4.3.5 U(VI) speciation in body fluids – TRLFS

4.3.5.1 Luminescence characteristics of U(VI) in saliva

As previously shown, speciation calculations revealed the $Ca_2UO_2(CO_3)_{3(aq)}$ as a prominent species at saliva pH range of 6 to 7 (Fig.4-22(a)) if the data about $Ca^{2+}/UO_2^{2+}/PO_4^{3-}$ system was not included. However, the fluorescence characteristics for both artificial and natural saliva did not indicate $Ca_2UO_2(CO_3)_{3(aq)}$ and/or U(VI)-phosphate species (refer to Fig.4-28 and Fig.4-29) which is uncommon in solutions contain relatively high carbonate as well as calcium ions. Moreover, the fluorescence bands positions were comparable to that of $Ca^{2+}/UO_2^{2+}/PO_4^{3-}$ system which was reported by Geipel (2005b) who studied that system but not in the presence of carbonate. In order to cross-check the speciation in both saliva fluids, two synthetic solutions were made and were further investigated by cryo-TRLFS. The first solution was consisted of 50 μM U(VI) and 10 mM PO_4^{3-} (UO_2^{2+}/PO_4^{3-} system). The second solution was composed of 50 μM U(VI), 10 mM PO_4^{3-}, and 1.2 mM Ca^{2+} ($Ca^{2+}/UO_2^{2+}/PO_4^{3-}$ system). The pH of both synthetic solutions was adjusted to 7, and then they were measured by cryo-TRLFS. Both solutions displayed very different fluorescence characteristics which means that the Ca^{2+} ions has a great influence on the UO_2^{2+}/PO_4^{3-} system. Then the fluorescence data of both systems were compared to that of artificial as well as natural saliva specimens (Fig.4-28).

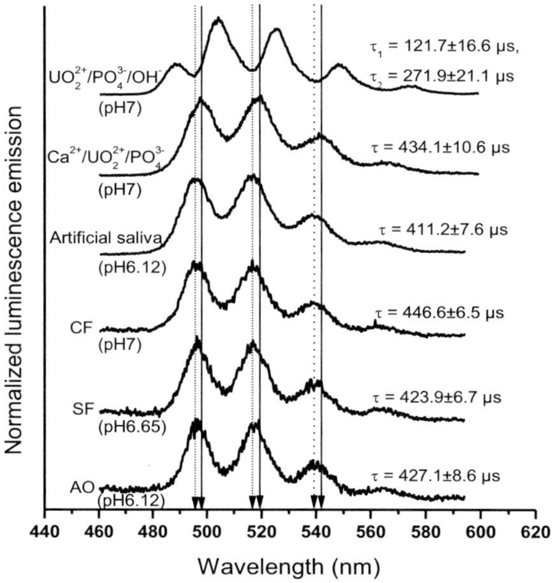

Figure 4-28. Fluorescence characteristics of U(VI) species in both natural (CF, SF, AO) and artificial saliva in comparison to some synthetic solutions at pH 6 – 7, T = -120°C, and λ_{em} = 266 nm

Referring to figure (4-28), we observe that original saliva specimens have almost the same fluorescence properties as $Ca^{2+}/UO_2^{2+}/PO_4^{3-}$ system and slightly different from artificial saliva and completely different from UO_2^{2+}/PO_4^{3-} system. Hence, $Ca^{2+}/UO_2^{2+}/PO_4^{3-}$ system seems important to characterize U(VI) speciation in original saliva as they have almost the same fluorescence properties. Geipel (2005b) managed to calculate the thermodynamic stability constants of the $Ca^{2+}/UO_2^{2+}/PO_4^{3-}$ system which have been used in our speciation calculations (refer to Eqs (4-9, 4-10)). Nevertheless, the fluorescence spectra of U(VI) binding forms in original saliva can also be compared to some reference luminescence data of U(VI)-phosphates and aqueous U(VI) complexes with biochemical compounds that commonly exist in human saliva (Table 4-18). For instance, the luminescence spectral bands of the pure α-amylase-U(VI) system over the pH range 4 to 7 are somewhat similar to that of natural saliva. This slight difference could be due to the complexity of the natural saliva and to the variation of α-amylase secretion and concentrations from that of the pure system. This degree of similarity is a bit consistent with the theoretical speciation when α-amylase was included in the equation model, as we have seen that the complex $UO_2[(COO)_2\text{-Amy}]^{2+}$ predominated until around pH 6.6. Another important aqueous complex that has almost typical fluorescence spectrum as natural saliva is *O*-phosphoserine-U(VI). However, it was not included in the speciation calculations since it is found in human saliva as part of a large protein (proline-rich protein) and not as a free molecule. Worth mentioning that at the biological pH range of human saliva (6 – 7) the luminescence characteristics of U(VI) species did not change (Fig.4-28 and Fig.4-29).

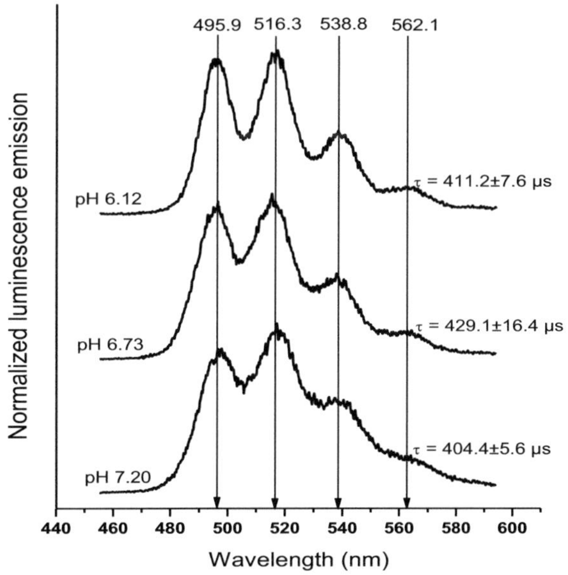

Figure 4-29. pH dependant experiment of U(VI) speciation in artificial saliva at $[UO_2^{2+}]$: 1 μM, T = -120°C, and λ_{em} = 266 nm

Table 4-18. Comparison of luminescence maxima for some uranium reference data with artificial and natural human saliva

Solution	Luminescence emission bands (nm)						Reference
UO_2^{2+}	486.6	508	531.1	556.8	584	-	This work
$UO_2H_2PO_4^+$	-	494	515	539	565	-	Scapolan et al. 1998a
UO_2HPO_4	-	497	519	543	570	-	Scapolan et al. 1998a
$UO_2PO_4^-$	-	499	520	544	571	-	Scapolan et al. 1998a
	-	502	524	548	574	-	Bonhoure et al. 2007
$(UO_2)_x(PO_4)_y$	488	503	524	547	573	601	Geipel et al. 2000
UO_2^{2+} (*O*-Phosphoserine)	482	496.2	516.5	539.2	563.7	593	Koban and Bernhard(2007)
UO_2^{2+} (α-Amylase) at pH 6 – 7	-	499	520	541	-	-	Barkleit and Hennig(2013)
UO_2^{2+} (α-Amylase) at pH 4 – 5.5	-	495	516	539	-	-	
$U(VI)/PO_4^{3-}/Ca^{2+}_{(aq)}$	-	497.8	518.8	540.7	565.3	-	This work
Artificial saliva	-	495.9	516.3	538.8	562.1	-	This work
Original saliva (AO)	-	496.5	517.6	539.5	563.5	-	This work
Original saliva (SF)	-	496.4	517.1	539	563.4	-	This work
Original saliva (CF)		496.0	516.7	538.6	562.7	-	This work

4.3.5.2 Luminescence characteristics of U(VI) in sweat

Referring to figure (4-30) luminescence characteristics of U(VI) in both artificial and natural sweat indicated the aqueous $Ca_2UO_2(CO_3)_3$ which is a typical binding form in natural waters. In other words, uranium in human sweat behaves like in natural waters and no real influence from the bioligands (urea, uric acid, lactic acid) that are secreted by sweat glands.

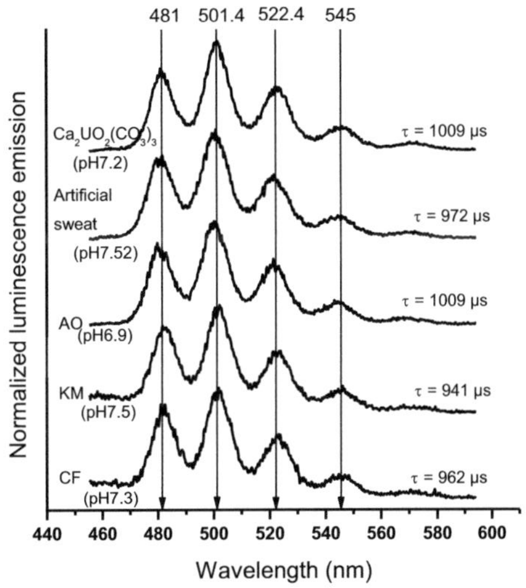

Figure 4-30. Fluorescence characteristics of U(VI) species in both natural (CF, KM, AO) and artificial sweat in comparison to some synthetic solutions at pH 6 – 8, T = -120°C, and λ_{em} = 266 nm

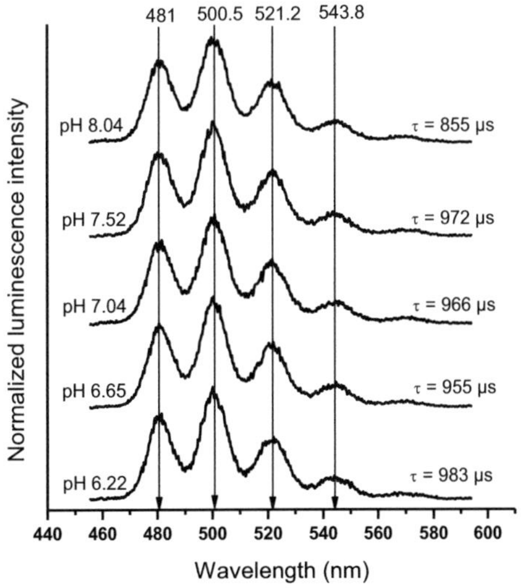

Figure 4-31. pH dependant experiment of U(VI) speciation in artificial sweat at $[UO_2^{2+}]$: 1 μM, T = -120°C, and λ_{em} = 266 nm

At the indicated pH range (6 to 8) fluorescence data of U(VI) species remains unchanged with only exception of artificial sweat at pH 8.04 (Fig.4-30 and Fig.4-31). The thermodynamic modeling approved the same U(VI) species as indicated by fluorescence measurement.

4.3.5.3 Luminescence characteristics of U(VI) in urine

In artificial urine, differences in U(VI) luminescence spectra were observed with alteration of pH values (Fig.4-32) and Ca^{2+} concentrations (Fig.4-33(b)). Here the system $Ca^{2+}/UO_2^{2+}/PO_4^{3-}$ is crucial in U(VI) distribution as the sparingly soluble complex $Ca(UO_2)_2(PO_4)_{2(aq)}$ can be formed depending on the pH, Ca^{2+} concentration, and time. This complex is constituted at pH ≤ 7.5 when Ca^{2+} concentration exceeds 2 mM and/or when the freshly prepared solution becomes older (aged). In the latter case it is assumed that at first the 1:1:1 complex ($CaUO_2PO_4^{+}$) is formed and then altered over time and formed the $Ca(UO_2)_2(PO_4)_{2(aq)}$ as a fine suspension (Fig.4-33(a)). This is somehow consistent with the speciation calculation of artificial urine since the $Ca(UO_2)_2(PO_4)_{2(aq)}$ prevailed until pH 7.5, after then the $Ca_2UO_2(CO_3)_{3(aq)}$ dominated the speciation (Fig.4-26 (b)).

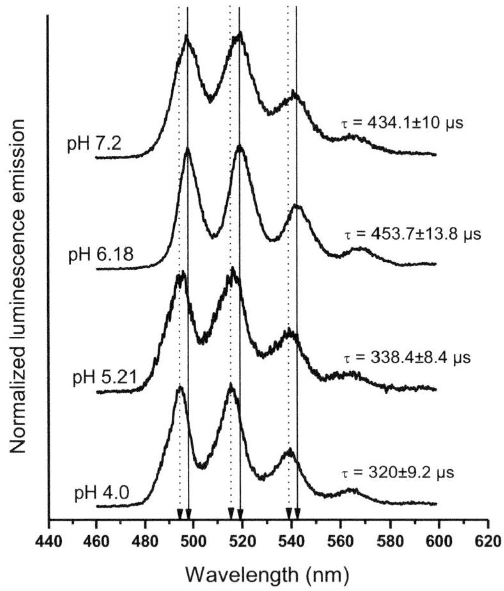

Figure 4-32. pH dependant experiment of U(VI) speciation in artificial urine at [UO_2^{2+}]: 50 µM, [PO_4^{3-}]: 12.3 mM, [Ca^{2+}] = 1.2 mM, T = -120°C, and λ_{em} = 266 nm

In contrast, U(VI) in natural human urine specimens under investigation (pH 5 – 6) showed different luminescence characteristics from that in artificial urine solution (Fig.4-34). This is mainly attributed to the matrix complexity of natural human urine (e.g. organics, ionic strength). The emission bands maxima of natural urine are comparable but not exactly similar to that of pure systems of UO_2^{2+}/Cit^{3-} and $UO_2(Cit)^-$ (Table 4-19). Speciation modeling (Fig.4-27) supported this assumption since it exhibited $UO_2(Cit)^-$ as the main binding form in the original urine specimens (pH range 5 – 6).

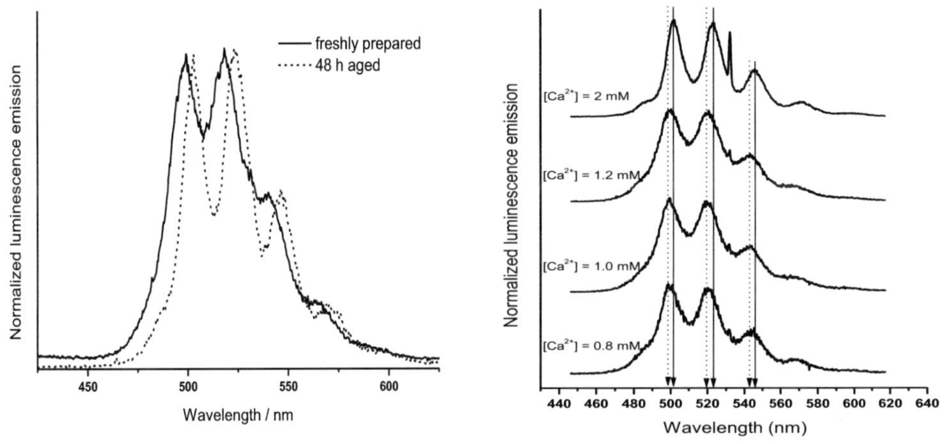

(a) (b)

Figure 4-33. Influence of (**a**) aged process, (**b**) increasing Ca^{2+} concentration on the system $Ca^{2+}/UO_2^{2+}/PO_4^{3-}$ at pH 7.2, $[UO_2^{2+}]$: 50 µM, $[PO_4^{3-}]$: 10 mM, T = -120°C, and λ_{em} = 266 nm.

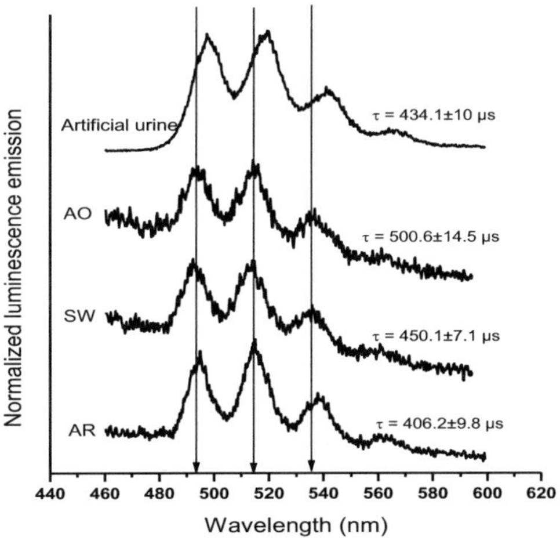

Figure 4-34. Fluorescence characteristics of U(VI) species in both natural (AR, SW, AO) and artificial human urine in comparison to some synthetic solutions at pH 5 – 7, T = -120°C, and λ_{em} = 266 nm

Table 4-19. Comparison of luminescence maxima for some uranium reference data with artificial and natural human urine

Solution	Luminescence emission bands (nm)						Reference
UO_2^{2+}	486.6	508	531.1	556.8	584	-	This work
$UO_2H_2PO_4^+$	-	494	515	539	566	-	Scapolan et al. 1998a
UO_2HPO_4	-	497	519	543	570	-	Scapolan et al. 1998a
$UO_2PO_4^-$	-	499	520	544	571	-	Scapolan et al. 1998a
	-	502	524	548	574	-	Bonhoure et al. 2007
$(UO_2)_x(PO_4)_y$	488	503	524	547	573	601	Geipel et al. 2000
UO_2HGly^{2+}	476	492	513	537	562	587	Guenther et al. 2007
$UO_2(HGly)_2^{2+}$	479	495	517	541	565	594	Guenther et al. 2007
$UO_2(Cit)^-$ (pH 2 – 4)	475.3	491.8	513.5	537	561.9	-	Guenther et al. 2011
$(UO_2)_2(Cit)_2^{2-}$ (pH 2 – 4)	483.6	502.7	524.5	548.1	574	-	Guenther et al. 2011
UO_2^{2+}/Cit^{3-} system (pH 5.55)	-	494.7	515.5	538.3	560.3	-	This work
$U(VI)/PO_4^{3-}/Ca^{2+}_{(aq)}$	-	497.8	518.8	540.7	565.3	-	This work
Artificial urine (pH 7.2)	-	497.8	518.4	540.1	565	-	This work
Original urine (AO)	-	494.1	514.4	536.7	562.9	-	This work
Original urine (SW)	-	493.2	513.6	535.4	561.1	-	This work
Original urine (AR)	-	494.8	515.2	537.8	562.4	-	This work

4.3.6 Concluding remarks

In the present work, the first experimental data on U(VI) speciation in natural body fluids (saliva, sweat, urine) was obtained by means of fluorescence spectroscopy. As a non-invasive technique for direct uranium speciation in body fluids, TRLFS showed a very good capability when coupled with cryogenic system which suppressed effect of fluorescence quenchers (organics, CO_3^{2-}, Cl^-) by cooling samples to very low temperatures (153 K). By the use of temporal and/or spectral information of the emission spectra, various uranyl complexes were demonstrated in the studied samples.

Variations of the body fluids in terms of chemical composition, pH, and ionic strength resulted in different binding forms of U(VI). Besides that, we elucidated the significant role of bioligands commonly exist in body fluids in U(VI) distribution at the biological pH range of each body fluid. A good example was the binding of U(VI) with α-amylase in saliva. Also citrate in urine seems an important complexant for U(VI), while lactate controlled speciation at pH ≤ 5.5 in sweat. But not all bioligands play such a role although they are found at relatively high concentrations like urea and uric acid in urine and sweat. An important point was the effect of phosphates and carbonates as biological buffers ($H_2PO_4^-/HPO_4^{2-}$, HCO_3^-/CO_3^{2-}) for U(VI) speciation in the body fluids especially in the presence of Ca^{2+} ions. For instance, in carbonate-poor urine (specimens AR and AO) the complex $Ca(UO_2)_2(PO_4)_{2(aq)}$ could give a new perspective for bioavailability of uranium within the renal system since it is sparingly soluble above pH 7.5 and can form a fine precipitate with the time.

To sum up, uranium binding forms have been demonstrated in *real natural body fluids* for the first time by both spectroscopy and thermodynamic modeling. Both methods indicate the important role of bioligands on U(VI) speciation which was not taken into account by previous studies. Although some thermodynamic data for aqueous uranium complexes were not compiled by the NEA database, chemical thermodynamic modeling provided a good estimation of U(VI) speciation in the body fluids which is somehow consistent with experimental data.

For more accurate calculations, the interaction of U(VI) with metabolites and biomolecules (proteins, lipids, etc) should be extended and reviewed for future investigation since it is not easy to assign all uranium species in body fluids by means of spectroscopy. However, application of cryo-TRLFS with extremely low temperature (< 4

K) with a direct excitation by a wavelength 410 to 460 nm will be promising as it enhances phosphorescence spectroscopy with elongated lifetime and a better bands resolution. Moreover, it offers the possibility for more detailed structural information about binding forms and coordination of uranyl ions. A combination of both spectroscopy and modeling can provide a comprehensive knowledge that helps for more insight research on uranium biokinetics. Other body fluids such as plasma, gastric juice, pancreatic juice should be included in future investigation to get a clear picture of the complicated metabolic pathways of uranium in the body after exposure.

4.4 Conclusion and Outlook

The main aim of this study is to find out the uranium speciation in environmentally relevant waters and in bio-fluids (saliva, sweat, and urine) by combining both theoretical modeling and spectroscopic methods. The chemical species $Ca_2UO_2(CO_3)_{3(aq)}$, which is known as a less toxic species (Carriere et al. 2004), was assigned to be the main binding form in mineral waters as well as in groundwater. The elevation in uranium concentration did not influence its binding forms at the pH range of drinking waters, whereas the increase in other water parameters such as Ca^{2+} and HCO_3^- changed the distribution greatly. The modeling results for the main uranium binding form in mineral waters and groundwater using thermodynamic data exist in the NEA database and some recent literature, indicated a good agreement with the spectroscopic measurements.

The obtained data could be used for more insight research on uranium chemo-toxicity in drinking water since there is no available information on the long-term exposure of low uranium concentrations via ingestion on humans. The only published data was the nephrotoxicity in rats after a short-term and a high dose uranium exposure (Carriere et al. 2005). Comparing to the study of Zamora et al. (1998) who focused on the total uranium concentration in drinking water and its implication on human health, this study offers the possibility to provide data on the binding forms. Therefore, it could be a baseline for epidemiological studies in the case of drinking waters of naturally high uranium levels such as in Pofadder arid area, South Africa. Moreover, such knowledge could also give a good perspective on uranium speciation in human body fluids after incorporation.

Uranium speciation in acid drainage waters (mining related waters, industrial wastewater) was affected by acidity and the presence of inorganic anions. The luminescence data of uranium in acid drainage waters reflect the importance of chloride ions even in the presence of other stronger ligands such as sulfate and phosphate. Speciation data of uranium in drainage waters, which is produced by human activities, is valuable for remediation strategies on contaminated sites (old uranium mines) and/or designing concept for waste repositories.

The results of complexation of U(VI) with some basic organic constituents (uric acid, urea) of human biological fluids were used in the speciation modeling of human body fluids. Taking into account such organic constituents, regardless of their weak complexation, in the equation model beside inorganic constituents contribute in a better understanding and/or accurate calculations and simulations of the intricate uranium speciation in human biological fluids. Moreover, the interaction of U(VI) with metabolites and biomolecules (proteins, lipids, etc.) should be extended and reviewed for future investigation since it is not easy to assign all uranium species in body fluids by means of spectroscopy.

The findings add also new data regarding the uranium(VI) speciation in human body fluids and elucidate the important role of bioligands commonly present in such fluids. The data provides a comprehensive knowledge and a better understanding as a basic mechanism of redistribution of uranium after exposure, and thus for more insight research on its biokinetics. Other body fluids such as plasma, gastric juice, pancreatic juice should be included in future investigation to get a clear picture of the complicated metabolic pathways of uranium in the body after exposure.

On the basis of cryo-TRLFS method, the direct determination of uranium speciation in animals' bio-fluids will give a new insight on the biochemical behavior of uranium. Furthermore, application of cryo-TRLFS with extremely low temperature (< 4 K) with a direct excitation by a wavelength 410 to 460 nm will be promising as it enhances phosphorescence spectroscopy with elongated lifetime and a better bands resolution. This also offers the possibility for more detailed structural information about binding forms and coordination of uranyl ions.

5. Experimental Details

5.1 Chemicals and solutions

All chemicals that have been used in preparation of human body fluids and in experiments of uranium(VI) complexation with bioligands are listed in Table (5-1). Ultra-pure CO_2-free Milli-Q water with a resistivity of 18.2 MΩ/cm (Milli-RO/Milli-Q-System, Millipore, France) was used for preparation of all solutions.

Table 5-1. The used chemical substances

Chemical substance	Formula	Purity	Producer
Urea	CH_4N_2O	99.5%	Acros Organics
Uric acid	$C_5H_4N_4O_3$	99%	VWR Prolabo
Sodium chloride	NaCl	99.99%	Merck
Sodium bicarbonate	$NaHCO_3$	99.998%	Alfa Aesar
Sodium sulfate	Na_2SO_4	p.a.	Merck
Sodium phosphate dibasic dodecahydrate	$Na_2HPO_4.12H_2O$	$\geq$ 99%	Fluka
Potassium bicarbonate	$KHCO_3$	99.7%	Sigma-Aldrich
Potassium phosphate monobasic	KH_2PO_4	$\geq$ 99.5%	Merck
Magnesium chloride hexahydrate	$MgCl_2.6H_2O$	p.a.	Merck
Magnesium sulfate heptahydrate	$MgSO_4.7H_2O$	p.a.	Merck
Calcium chloride dihydrate	$CaCl_2.2H_2O$	p.a.	Merck

5.1.1 Uranium(VI) stock solution

0.1 M $UO_2(ClO_4)_2$ was prepared by dissolving the appropriate amount of UO_3 in 0.3 M $HClO_4$. A working stock solution of $UO_2(ClO_4)_2$ (5×10^{-3} M in 0.02 M $HClO_4$) was prepared by dilution from the main stock solution (0.1 M $UO_2(ClO_4)_2$). Exact concentrations for both solutions were determined by ICP-MS.

5.1.2 Complexation experiments of U(VI) with urea and uric acid

Exact weights of 1.502 g urea and 0.0042 g uric acid were dissolved in 25 mL Milli-Q water without further purification to prepare 1 M and 1 mM stock solutions, respectively.

A working stock solution of $UO_2(ClO_4)_2$ (5×10^{-3} M in 0.02 M $HClO_4$), whose exact concentration was determined by ICP-MS, was used. All complexation experiments were performed at a fixed uranyl concentration of 5×10^{-5} M and various ligand concentrations ranged from 8×10^{-6} M – 10^{-4} M for uric acid and 10^{-5} M – 10^{-4} M, 0.01 M – 0.3 M for two sets of urea. Ionic strength was adjusted to 0.1 M $NaClO_4$. The pH value was kept constant at 3 for U(VI) – uric acid system and 4 in the case of U(VI) – urea system.

5.1.3 Preparation of artificial body fluids

Artificial body fluids were prepared by dissolving chemical salts that were accurately weighed in a 500 mL volumetric flask contained 250 mL carbonate-free Milli-Q water. After complete dissolution of the salts the volume was then completed to the mark with the same type of water. Then the constituents of the dissolved salts were checked by elemental analysis techniques, ICP-MS for cations and ion chromatography for anions.

Table 5-2. Amounts of chemical salts used in preparation of artificial human saliva

Salt	**Amount (g/500 mL)**	**Salt**	**Amount (g/500 mL)**
$MgCl_2.6H_2O$	0.0202	KH_2PO_4	0.3999
$CaCl_2.2H_2O$	0.1107	Na_2HPO_4	0.1330
NaCl	0.2999	$KHCO_3$	0.287

Table 5-3. Amounts of chemical salts used in preparation of artificial human sweat

Salt	**Amount (g/500 mL)**	**Salt**	**Amount (g/500 mL)**
NaCl	2.0008	$KHCO_3$	0.503
$CaCl_2$	0.0598	$MgSO_4$	0.0486

Table 5-4. Amounts of chemical salts used in preparation of artificial human urine

Salt	**Amount (g/500 mL)**	**Salt**	**Amount (g/500 mL)**
NaCl	2.298	KCl	2.201
$CaCl_2.2H_2O$	0.088	Na_2SO_4	1.156
$NaHCO_3$	0.311	$MgSO_4.7H_2O$	0.306
KH_2PO_4	0.923		

5.2 Methods

5.2.1 SPECFIT

The multivariate factor analysis program SPECFIT/32 (Binstead et al. 2005) was also used to calculate the stability constants (log β) of U(VI) – uric acid and U(VI) – urea systems. This program depends on fitting a variety of different complex spectra with multiple components, to calculate the protonation and stability constants as well as single component spectra. It functions on the basis of principal component analysis, in which an eigen value decomposition (Single Value Decomposition – SVD) is performed. Input parameters for the program include concentrations and dissociation constants of the substances involved in the reaction and the pH of each sample. In addition, it is possible to specify precisely defined and known spectra as starting point for the fit. The fit is performed on the chi-square minimization using different algorithms (e.g. Marquardt, Gridfit, Alternate). Depending on the relationship between the concentrations of different species according to equilibrium and a given fit model; the program calculates the stoichiometry of the formed complexes of U(VI) and their stability constants. Gampp et al. (1985) reported descriptive details of SPECFIT and its applications in equilibrium constants calculations.

5.2.2 Thermodynamic modeling

Speciation calculations for U(VI) in natural waters were performed using the geochemical code EQ3/6 (Wolery 1992). Calculations by EQ3/6 are always good for solutions with ionic strength $\leq$ 0.1 M. Therefore, due to high ionic strength in some body fluids such as urine (I: 0.3 – 0.7 M) we preferred to use another code with more flexibility with higher ionic strength. For this purpose the computer codes HYDRA and MEDUSA (Puigdomenech 2004) were used. HYDRA contains a database with all basic parameters including equilibrium constants that are necessary for the calculation, and MEDUSA creates many kinds of diagrams. In the current research, calculations were carried out by both codes using the most recent thermodynamic stability constants for aqueous uranium(VI) complexes that are compiled in the OECD/NEA database (Guillaumont et al. 2003). However, other non-compiled data was also used in our calculations as detailed in the section "Results and Discussion".

5.2.3 Elemental analysis

Compositional anaylsis (cations and anions) for both environmental waters and body fluids was performed by inductively coupled plasma – mass spectrometry (ICP-MS) and ion chromatography (IC). Uranium and main cations were determined by using ICP-MS (Elan 9000 from Perkin Elmer, Boston, USA), and the main anions were determined by using ion chromatography (IC-System 732/733 from Metrohm, Switzerland).

5.2.4 Determination of carbon

Organic composition was quantified as a total organic carbon (TOC) and HCO_3^-/CO_3^{2-} as a total inorganic carbon (TIC) by using a total carbon analyzer (Multi N/C 2100 from Analytik Jena AG, Jena, Germany).

5.2.5 Measurement of pH and EC values

The pH was measured using a glass probe, Blue Line 16 pH (Schott Instruments, Germany) calibrated with standard buffer solutions (pH 4.00 and pH 6.86). The pH meter, WTW inoLab pH 720 (WTW, Weilheim, Germany) had an accuracy of ±0.01 pH units. Electrical conductivity (EC) was measured using a LF 325 conductivity meter with TetraCon® 325 Standard-Conductivity cell (WTW, Weilheim, Germany).

5.3 Spectroscopy technique – TRLFS

Time-resolved laser-induced fluorescence spectroscopy (TRLFS) is the methodical basis throughout the whole current research work. It is known as a unique, rapid, and sensitive technique for direct uranium speciation in environmentally relevant concentrations. The principle is based on the excitation of a uranyl ion by a suitable wavelength, and the recorded spectrum is observed by the simultaneous detection of peaks positions and the lifetime of a fluorescent light. By comparing these spectroscopic characteristics with the corresponding spectra of known uranium compounds it is possible to identify the unknown species. Worth to mention is that the emission is normally observed at a 90° angle from the incident laser beam to separate it from transmitted light.

5.3.1 TRLFS measurements of U(VI) complexes with urea and uric acid

All spectroscopic experiments were carried out with a Nd:YAG laser system (Continuum Minilite II, Continuum Electro Optics Inc., USA). A laser wavelength at 266 nm with pulse energy of 0.3 mJ was used for exciting the uranyl luminescence. The trigger input of the detection system was coupled directly to the trigger output of the laser. The luminescence signal was focused into the spectrograph (iHR 550, HORIBA JOBIN YVON GmbH, Germany) by means of an optical fiber cable, and then detected by an ICCD camera system with 1024 useable pixels and built in a delay generator (HORIBA Jobin Yvon). The whole functions and setting of the laser spectrometer were controlled by the computer software LabSpec5 (HORIBA Jobin Yvon). The TRLFS spectra were recorded in the range from 371.39 to 674.28 nm averaging three time-resolved spectra with 50 laser shots for each single spectrum. The gate time was set to 2 μs, 20 μs, and the delay varied from 150 – 10000 ns, 50 - 40000 ns for uric acid and urea, respectively.

5.3.2 Cryo-TRLFS measurements for natural waters and body fluids

For both natural waters and body fluids, TRLFS measurement at a low temperature (153 K) was adopted since it minimizes the fluorescence quench effects and improves thc detection limit and makes it much lower (~10^{-10} M) than at room temperature (~10^{-7} M). All measurements were performed using a Nd:YAG laser system (Inlite Continuum®, Electro-optics, Inc. USA) operating at 266 nm delivering an energy of 5 mJ in a 10 ns pulse, and with a repetition rate of 20 Hz. Unlike other wavelengths, 266 nm has higher excitation energy and excites all uranyl compounds non-specifically. The emitted fluorescence light was focused by an optical fiber into a spectrograph (Acton SpectraPro® 300i, Göttingen, Germany) and collected by an intensified camera system (ICCD from Princeton Instrument, Inc. USA). The whole spectrometer and its functions were computer controlled using the data acquisition software WinSpec32 (Princeton Instrument, Inc. USA). For this laser setup, a cryogenic system (KGW-ISOTHERM, Typ SC2, Karlsruhe, Germany) was attached. This system controls the temperature at 153 K (-120°C) via a constant flow of nitrogen gas. A typical setup for cryogenic TRLFS is shown in figure (5-1). Other measurements parameters for both environmental waters and body fluids are shown in Tables (5-5 & 5-6).

Table 5-5. TRLFS – Measurement parameters for different environmental waters

Mineral waters	**Groundwater**	**Acid drainage waters**
Gate width = 500 μS Delay time = 4 mS Delay increment = 40 μS No. of spectra = 101 No. of laser shots = 300 Accumulations = 3	Gate width = 100 μS Delay time = 4 mS Delay increment = 40 μS No. of spectra = 101 No. of laser shots = 30 Accumulations = 3	Gate width = 50 μS Delay time = 1.5 mS Delay increment = 15 μS No. of spectra = 101 No. of laser shots = 40 Accumulations = 3

Table 5-6. TRLFS – Measurement parameters for different body fluids

Saliva	Sweat	Urine
Gate width = 50 μS Delay time = 1.5 mS Delay increment = 15 μS No. of spectra = 101 No. of laser shots = 50 Accumulations = 3	Gate width = 50 μS Delay time = 4 mS Delay increment = 40 μS No. of spectra = 101 No. of laser shots = 50 Accumulations = 3	Gate width = 50 μS Delay time = 1.5 mS Delay increment = 15 μS No. of spectra = 101 No. of laser shots = 50 Accumulations = 3

5.3.3 Sample preparation for cryo-TRLFS measurement

Small aliquot (3 mL) of the sample (either water or a bio-fluid) was taken and transferred into a plastic cuvette (PLASTIBRAND, BRAND – Wertheim, Germany), closed, and shock frozen using liquid nitrogen and then stored in a deep freezer (-20°C). This frozen sample was eventually extracted from the plastic cuvette and taken as an ice block and placed in a specially designed holder (see Fig.5-2) and then put inside the above indicated cryogenic system. The sample was let for 30 minutes in the cooling system before running the TRLFS measurement to reach the required temperature of 153 K (heat equilibrium).

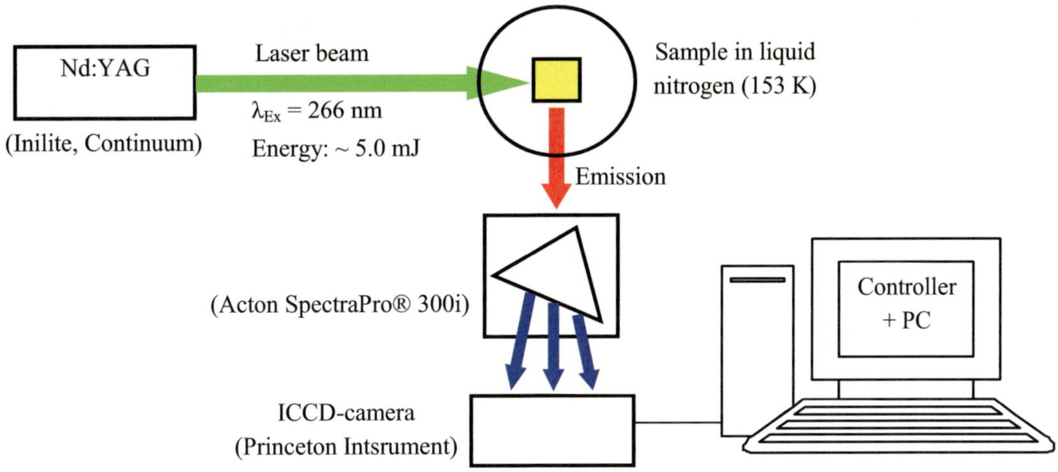

Figure 5-1. A typical setup of a cryogenic time-resolve laser-induced fluorescence spectroscopy dedicated for uranium speciation in natural waters and body fluids

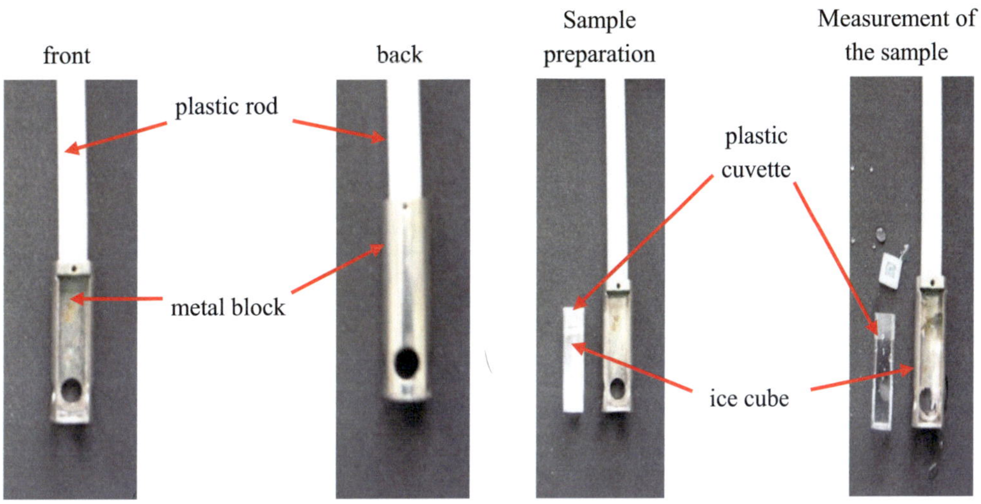

Figure 5-2. A specially designed sample holder to measure ice block samples in the frozen state (153 K)

5.4 Data on mineral water samples

The mineral waters under investigation were from different German producers. They were bought from local supermarkets in Lower Saxony, Bavaria, Hesse, Thuringia, Baden-Württemberg, and the North Rhine-Westphalia States. Detailed information on water type, sources, and analysis that is provided by the producers is in Tables (5-7 & 5-8).

Table 5-7. Summary analysis of the main inorganic constituents (in mg/L) of the investigated mineral waters (according to the producers)

Sample ID	Na^+	Mg^{2+}	Ca^{2+}	F^-	Cl^-	SO_4^{2-}	HCO_3^-
Extalerquelle	12	60.5	353	-	-	890	256
Nestle Aquarel	38	29	108	-	52	0	463
Thüringer Waldquell	22.4	39.6	84.3	-	36.2	161	230
Adelholzener	12.9	31.3	67.1	0.17	20.5	27	319
Alwa	17	65	485	-	28	1110	412
Teinacher	73.8	26.7	96.2	-	15.8	26	567
Warburger Waldquell	28.2	50.7	263	1.1	16	649	353.3

Table 5-8. Classification of the investigated mineral waters, along with the brands, sources, and date of sampling

Type of water	Brand	Source	State	Sampling date
STILL	Extalerquelle	Extalerquelle	Lower Saxony	16.03.2011
STILL	Nestle Aquarel	Birken-Quelle	Hesse	14.03.2011
STILL	Thüringer Waldquell	Schmalkalden	Thuringia	22.06.2011
CLASSIC	Adelholzener	Allpenquellen	Bavaria	19.10.2011
CLASSIC	Alwa	Alwa-Quelle	Baden-Württemberg	17.11.2011
CLASSIC	Teinacher	Bad Teinach	Baden-Württemberg	27.10.2011
CLASSIC	Warburger Waldquell	Different sources	North Rhine-Westphalia	15.11.2011

6. References

Albani JR (2008) Principles and Applications of Fluorescence Spectroscopy. Blackwell Publishing Ltd, Oxford, UK

ATSDR (Agency for Toxic Substances and Disease Registry) (1999) Toxicological profile for uranium. U.S. Department of Health and Human Services. Public Health Service, Atlanta, GA

Balzani V, Boletta F, Gandolfi MT, Maestri M (1978) Bimolecular Electron Transfer of Excited States of Transition Metal Complexes. Top Curr Chem 75:1-64

Barkleit A, Hennig C (2013) Interaction of the enzyme alpha-amylase with UO_2^{2+}. Wissenschaflich-technische Berichte, HZDR-048, p20

Bell JT, Biggers RE (1968) Absorption spectrum of the uranyl ion in perchlorate media III. J Mol Spectrosc 25:312-329

Bergmann E, Weizmann A (1938) The structure of Urea and its Derivatives. Trans Faraday Soc 34:783-786

Berlin M, Rudell B (1986) Uranium. In: Friberg L, Nordberg GF, Vouk VB (ed) Handbook on the toxicology of metals. Elsevier Science Publishers, Amsterdam, pp. 623-637

Bernhard G, Geipel G, Brendler V, Nitsche H (1996) Speciation of uranium in seepage waters of a minetailing pile studied by time-resolved laser-induced fluorescence spectroscopy (TRLFS). Radiochim Acta 74:87-91

Bernhard G, Geipel G, Reich T, Brendler V, Amayri S, Nitsche H (2001) Uranyl(VI) carbonate complex formation: Validation of the $Ca_2UO_2(CO_3)_{3(aq.)}$ species. Radiochim Acta 89:511-518

BfR (Bundesanstalt für Risikobewertung) und BfS (Bundesanstalt für Strahlenschutz) (2007) BfR empfiehlt die Ableitung eines europäischen Höchstwertes für Uran in Trink- und Mineralwasser. Gemeinsame Stellungnahme Nr. 020/2007 des BfS und des BfR vom 5. April 2007

Billard I, Geipel G (2008) Luminescence analysis of actinides: instrumentation, applications, quantification, future trends, and quality assurance. In: Resch-Genger U (ed) Standardization and Quality Assurance in Fluorescence Measurements I. Springer-Verlag, Berlin, pp 465-492

Biltz H, Herrmann L (1921) Acidität der Wasserstoffatome in der Harnsäure. Berichte der Deutschen Chemischen Gesellschaft/ B, pp 1676-1694

Binstead RA, Zuberbühler AD, Jung B (2005) SPECFIT – Global Analysis System, Version 3.0.37. Spectrum Software Associates, Marlborough, USA

Bonhoure I, Meca S, Marti V, De Pablo J, Cortina JL (2007) A new time-resolved laser-induced fluorescence spectrometry (TRLFS) data acquisition procedure applied to the uranyl-phosphate system. Radiochim Acta 95:165-172

Bowman FJ, Foulkes EC (1970) Effects of Uranium on Rabbit Renal Tubules. Toxicol Appl Pharmacol 16:391-399

Brachmann A, Geipel G, Bernhard G, Nitsche H (2002) Study of uranyl(VI) malonate complexation by time resolved laser-induced fluorescence spectroscopy (TRLFS). Radiochim Acta 90:147-153

Brendler V, Geipel G, Bernhard G, Nitsche H (1996) Complexation in the System $UO_2^{2+}/PO_4^{3-}/OH^-_{(aq)}$: Potentiometric and Spectroscopic Investigations at very Low Ionic Strengths. Radiochim Acta 74:75-80

Brina R, Miller AG (1992) Direct detection of trace levels of uranium by laser-induced phosphorimetry. Anal Chem 64:1413-1418

Burrows HD, Pedrosa de Jesus JD (1976) A Flash Photolytic Study of the Photo-oxidation of some Inorganic Anions by the Uranyl Ion. J Photochem 5:265-275

Carriere M, Avoscan L, Collins R, Carrot F, Khodja H, Ansoborlo E, Gouget B (2004) Influence of uranium speciation on normal rat kidney (NRK-52E) proximal cell cytotoxicity. Chem Res Toxicol 17:446-452

Carriere M, Khodja H, Avoscan L, Carrot F, Gouget B (2005) Uranium(VI) complexation in cell culture medium: influence of speciation on normal rat kidney (NRK-52E) cell accumulation. Radiochim Acta 93:691-697

Choppin GR and Du M (1992) *f*-Element Complexation in Brine Solutions. Radiochim Acta 58/59:101-104

Clark DL, Keogh DW, Neu MP, Runde W (1997) Uranium and uranium compounds. In: Kroschwitz JI, Howe-Grant M (ed) Kirk-Othmer Encyclopedia of Chemical Technology. John Wiley & Sons, New York, pp 638-694

Cordfunke EHP (1969) The Chemistry of Uranium. Elsevier, Amsterdam

Cotton S (2006) Coordination Chemistry of the Actinides. In Lanthanide and Actinide Chemistry. John Wiley & Sons, Ltd, Chichester, pp 173-199

Draelos ZD (2010) Prevention of cosmetic problems. In: Norman RA (ed) Preventive dermatology. Springer-Verlag, London Limited, pp 173-186

Davies KJ, Sevanian A, Muakkassah-Kelly SF, Hochstein P (1986) Uric acid – iron complexes, A new aspect of the antioxidant functions of uric acid. Biochem J 235:747-754

Derbyshire PJ, Barr H, Davis F, Higson SPJ (2012) Lactate in human sweat: a critical review of research to the present day. J Physiol Sci 62:429-440

Dong W, Brooks SC (2006) Determination of the formation constants of ternary complexes of uranyl and carbonate with alkaline earth metals (Mg^{2+}, Ca^{2+}, Sr^{2+}, and Ba^{2+}) using anion exchange method. Environ Sci Technol 40:4689-4695

Bundesgesetzblatt Jahrgang 2011 Teil Nr. 21, ausgegeben zu Bonn am 11, Mai 2011. Erste Verordnung zur Änderung der Trinkwasserverordnung vom 3. Mai 2011

Eliet V, Bidoglio G, Omenetto N, Parma L, Grenthe T (1995) Characterisation of Hydroxide Complexes of Uranium(VI) by Time-resolved Fluorescence Spectroscopy. J Chem Soc, Faraday Trans 91:2275-2285

Fisenne IM, Perry PM, Decker KM, Keller HW (1987) The daily intake of $^{235,235,238}U$, $^{228,230,232}Th$ and $^{226,228}Ra$ by New York City residents. Health Phys 53:357-363

Fisenne IM, Perry PM, Harley NH (1988) Uranium in humans. Radiat Prot Dosim 24:127-131

Formosinho SJ, Miguel MD, Graca M (1984) Photophysics of the Excited Uranyl Ion in Aqueous Solution: Part 3. Effects on Temperature and Deuterated Water: Mechanisms of

Solvent Exchange and Hydrogen Abstraction from Water. J Chem Soc, Faraday Trans 80:1745-1756

Frengstad B, Kjersti A, Skrede M, Banks D, Krog JR, Siewers U (2000) The chemistry of Norwegian groundwaters: III. The distribution of trace elements in 476 crystalline bedrock groundwaters, as analysed by ICP-MS techniques. Sci Total Environ 246:21-40

Galletti M, D'Annibale L, Pinto V, Cremisini C (2003) Uranium daily intake and urinary excretion: a preliminary study in Italy. Health Phys 85:228-235

Gampp H, Maeder M, Zuberbühler AD (1985) Calculations of equilibrium constants from multiwavelength spectroscopic data – II32, 95.: SPECFIT: two user-friendly programs in BASIC and standard FORTRAN 77. Talanta 32:257-264

Geipel G, Brachmann A, Brendler V, Bernhard G, Nitsche H (1996) Uranium(VI) sulfate complexation studied by time resolved laser-induced fluorescence spectroscopy Radiochim Acta 75:199-204

Geipel G, Bernhard G, Brendler V, Reich T (2000) NRC5, in: Proceedings of the 5th International Conference on Nuclear and Radiochemistry, Extended Abstracts, Pontresina, Switzerland, 3 – 8 September, 2000, pp 473

Geipel G (2005a) Speciation of actinides. In: Cornelis R, Caruso J, Crews H, Heumann K (ed) Handbook of Elemental Speciation II: Species in the Environment, Food, Medicine, and Occupational Health. John Wiley & Sons, West Sussex, pp 509-563

Geipel, G (2005b) International Chemical Congress of Pacific Basin Societies (PACIFICHEM), Honolulu, Hawaii, USA, Dec. 15. – 20. (2005). In C&EN ARCHIVES, Chem Eng News, 83:56-61, DOI: 10.1021/cen-v083n041

Geipel G (2006) Laser-induced fluorescence spectroscopy. In: Vij DR (ed) Handbook of applied solid state spectroscopy. Springer, USA, pp 577-593

Geipel G, Amayri S, Bernhard G (2008) Mixed complexes of alkaline earth uranyl carbonates: A laser-induced time-resolved fluorescence spectroscopic study. Spectrochim Acta, Part A 71:53-58

Gilman AP, Villeneuve DC, Secours VE, Yagminas AP, Tracy BL, Quinn JM, Valli VE, Willes RJ, Moss MA (1998a) Uranyl nitrate: 28 day and 91 toxicity studies in the Sprague Dawley Rat. Toxicol Sci 41:117-128

Gilman AP, Moss MA, Villeneuve DC, Secours VE, Yagminas AP, Tracy BL, Quinn JM, Long J, Valli VE, Willes RJ. Uranyl nitrate (1998b) 91-day exposure and recovery studies in the New Zealand white rabbit. Toxicol Sci 41:138-151

Götz C, Geipel G, Bernhard G (2011) The influence of the temperature on the carbonate complexation of uranium(VI): a spectroscopic study. J Radioanal Nucl Chem 287:961-969

Grenthe I, Fuger J, Konings RJM, Lemire RJ, Mueller AB, Nguyen-Trung C, Wanner H (1992) Chemical Thermodynamics of Uranium. Elsevier, Amsterdam

Guenther A, Geipel G, Bernhard G (2007) Complex formation of uranium(VI) with the amino acids l-glycine and l-cysteine: A fluorescence emission and UV–Vis absorption study. Polyhedron 26:59-65

Guenther A, Steudtner R, Schmeide K, Bernhard G (2011) Luminescence properties of uranium(VI) citrate and uranium(VI) oxalate species and their application in the determination of complex formation constants. Radiochim Acta 99:535-541

Guillaumont R, Fanghaenel T, Neck V, Fuger J, Palmer DA, Grenthe I, Rand MH (2003) Chemical thermodynamics 5, Update on the Chemical Thermodynamics of Uranium, Neptunium, Plutonium, Americium and Technetium. Elsevier, Amsterdam

Guyton AC, Hall JE (2000) Text Book of Medical Physiology – 10^{th}Edition. WB Saunders Company, Philadelphia

Harley JH (1988) Naturally occurring sources of radioactive contamination. In: Harley JH, Schmidt GD, Silini G (ed) Radionuclides in the food chain. Springer-Verlag, Berlin, pp 58-71

Harris JO, Robson AH (1948) Structure of Urea. Nature 161:98-98

Health Canada (2009) Guidelines for Canadian Drinking Water Quality: Guideline Technical Document – Radiological Parameters. Radiation Protection Bureau, Healthy Environments and Consumer Safety Branch, Health Canada, Ottawa, Ontario (Catalogue No. H128-1/10-614E-PDF)

Heller A, Barkleit A, Bernhard G, Ackermann J (2009) Complexation study of europium(III) and curium(III) with urea in aqueous solution investigated by time-resolved laser-induced fluorescence spectroscopy. Inorg Chim Acta 362:1215-1222

Heller A (2011) Spektroskopische Untersuchungen zur Komplexbildung von Cm(III) und Eu(III) mit organischen Modellliganden sowie ihrer chemischen Bindungsform in menschlichem Urin (in vitro). PhD Thesis, Technical University of Dresden, Germany

Hennig C, Tutschku J, Rossberg A, Bernhard G, Scheinost AC (2005) Comparative EXAFS investigation of uranium(VI) and -(IV) aquo chloro complexes in solution using a newly developed spectroelectrochemical cell Inorg Chem 44:6655-6661

Hoebler C, Karinthi A, Devaux MF, Guillon F, Gallant DJG, Bouchet B, Melegari C, Barry JL (1998) Physical and chemical transformations of cereal food during oral digestion in human subjects. Br J Nutr 80:429-436

Huang C, Chen M, Huang L, Mao I (2002) Uric acid and Urea in Human Sweat. Chin J Physiol 45:109-115

Hursh JB, Spoor NL (1973) Data on man. In: Hodge HC, Stannard JN, Hursh JB (ed) Uranium, plutonium, transplutonic elements – Handbook of experimental pharmacology 36. Springer-Verlag, Berlin, pp 197-240

Iyengar GV, Kollmer WE, Bowen, HJM (1978) The Elemental Composition of Human Tissues and Body Fluids: a compilation of values for adults. Verlag Chemie, Weinheim Newyork

Karpas Z, Paz-Tal O, Lorber A, Salonen L, Komulainen H, Auvinen A, Saha H, Kurttio P (2005) Urine, hair, and nails as indicators for ingestion of uranium in drinking water. Health Phys 88:229-242

Kato Y, Meinrath G, Kimura T, Yoshida Z (1994) A Study of U(VI) Hydrolysis and Carbonate Complexation by Time-resolved Laser-induced Fluorescence Spectroscopy (TRLFS). Radiochim Acta 64:107-111

Kelly SD, Kemner KM, Brooks SC (2007) X-ray absorption spectroscopy identifies calcium-uranyl-carbonate complexes at environmental concentrations. Geochim Cosmochim Acta 71:821–834

Koban A, Geipel G, Roßberg A, Bernhard G (2004) Uranium(VI) complexes with sugar phosphates in aqueous solution. Radiochim Acta 92:903-908

Koban A, Bernhard G (2007) Uranium(VI) complexes with phospholipid model compounds - A laser spectroscopic study. J Inor Biochem 101:750-757

Kropp JL (1967) Energy Transfer in Solution between UO_2^{2+} and Eu^{3+}. J Chem Phys 46:843-847

Kurttio P, Komulainen H, Leino A, Salonen L, Auvinen A, Saha H (2005) Bone as a possible target of chemical toxicity of natural uranium in drinking water. Environ Health Perspect 113:68-72

Kurttio P, Harmoinen A, Saha H, Salonen L, Karpas Z, Komulainen H, Auvinen A (2006) Kidney toxicity of ingested uranium from drinking water. Am J Kidney Dis 47:972-982

Lakowicz JR (2006) Principles of fluorescence spectroscopy, 3rd Edition. Plenum, New York

Langmuir D (1997) Aqueous Environmental Geochemistry. Prentice Hall, New Jersey

Leggett RW (1989) The behaviour and chemical toxicity of U in the kidney: a reassessment. Health Phys 57:365-383

Lehmann S, Geipel G, Grambole G, Bernhard G (2009) A novel time-resolved laser fluorescence spectroscopy system for research on complexation of uranium (IV). Spectrochim Acta, Part A 73:902-908

Lichtner PC, Waber N (1992) Redox front geochemistry and weathering: theory with application to the Osamu Utsumi uranium mine, Poços de Caldas, Brazil. J Geochem Explor 45:521-564

Liu L, Mo H, Wei S, Raftery D (2012) Quantitative analysis of urea in human urine and serum by 1H nuclear magnetic resonance. Analyst 137:595-600

Lotnik SV, Khamidullina LA, Kazakov VP (2003a) Influence of temperature on the lifetime of electronically excited uranyl ion: I. Liquid and supercooled H_2SO_4 solutions. Radiochemistry 45:550-554

Lotnik SV, Khamidullina LA, Kazakov VP (2003b) Influence of temperature on the lifetime of electronically excited uranyl ion: II. Frozen aqueous solutions of H_2SO_4. Effect of phase transitions. Radiochemistry 45:499-502

Luetke L, Moll H, Bernhard G (2012) A new uranyl benzoate species characterized by different spectroscopic techniques. Radiochim Acta 100:297-303

Mandel AL, Peyrot des Gachons C, Plank KL, Alarcon S, Breslin PAS (2010) Individual differences in AMY1 gene copy number, salivary α-amylase levels, and the perception of oral starch. PLoS One 5:13352

Marcantonatos MD (1980) Photochemistry and Exciplex of the Uranyl Ion in Aqueous Solution. J Chem Soc, Faraday Trans 76:1093-1115

Markich SJ (2002) Uranium speciation and bioavailability in aquatic systems: an overview. Sci World J 2:707-729

Marques MRC, Loebenberg R, Almukainzi M (2011) Simulated biological fluids with possible application in dissolution testing. Dissolution Technol 18:15-28

Marsh W, Fingerhut B, Miller H (1965) Automated and manual direct methods for the determination of blood urea. Clin Chem 11:624-627

Matsushima R, Fujimori H, Sakuraba S (1974) Quenching of the Uranyl (UO_2^{2+}) Emission by Inorganic Ions in Solution. J Chem Soc, Faraday Trans 70:1702-1709

Meinrath G (1998) Aquatic chemistry of uranium – a review focusing on aspects of environmental chemistry. FOG, ISSN 1434-7512 (http://www.geo.tu-freiberg.de/fog), 1:1-99

Meinrath A, Schneider P, Meinrath G (2003) Uranium ores and depleted uranium in the environment, with a reference to uranium in the biosphere from the Erzgebirge/Sachsen, Germany. J Environ Radioact 64:175-193

Meyer F, Laitano O, Bar-Or O, McDougall D, Heigenhauser GJF (2007) Effect of age and gender on sweat lactate and ammonia concentrations during exercise in the heat. Braz J Med Biol Res 40:135-143

Moawad MM (2002) Complexation and thermal studies of uric acid with some divalent and trivalent metal ions of biological interest in the solid state. J Coord Chem 55:61-78

Moll H, Reich T, Hennig C, Rossberg A, Szabo Z, Grenthe I (2000) Solution coordination chemistry of uranium in the binary UO_2^{2+}-SO_4^{2-} and the ternary UO_2^{2+}-SO_4^{2-}-OH^- system Radiochim Acta 88:559-566

Montavon G, Apostolidis C, Bruchertseifer F, Repinc U, Morgenstern A (2009) Spectroscopic study of the interaction of U(VI) with transferrin and albumin for speciation of U(VI) under blood serum conditions. J Inorg Biochem 103:1609-1616

Moriyasu M, Yokoyama Y, Ikeda S (1977) Anion Coordination to Uranyl Ion and the Luminescence Lifetime of the Uranyl Complex. J Inorg Nucl Chem 39:2199-2203

Nair S, Merkel BJ (2011) Impact of Alkaline Earth Metals on Aqueous Speciation of Uranium(VI) and Sorption on Quartz. Aquat Geochem 17:209-219

NCRP (National Council on Radiation Protection and Measurements) (1999) Biological Effects and Exposure Limits for “Hot Particles”. NCRP Report No. 130, Bethesda, Maryland

Noble RE (2000) Salivary alpha-amylase and lysozyme levels: a non-invasive technique for measuring parotid vs. submandibular/sublingual gland activity. J Oral Sci 42:83-6

Oppenheim FG, Salih E, Siqueira WL, Zhang W, Helmerhorst EJ (2007) Salivary proteome and its genetic polymorphisms. Ann NY Acad Sci 1098:22-50

Osman AAA, Geipel G (2012) Luminescence properties of uranyl ions in chloride media in frozen samples. Wissenschaflich-technische Berichte, HZDR-030, pp 60

Osman AAA, Geipel G, Bernhard G, Worch E (2013a) Investigation of uranium binding forms in selected German mineral waters. Environ Sci Pollut Res 20:8629-8635

Osman AAA, Geipel G, Bernhard G (2013b) Interaction of uranium(VI) with bioligands present in human biological fluids: the case study of urea and uric acid. Radiochim Acta 101:139-148

Ozutsumi K, Taguchi Y, Kawashima T (1995) Thermodynamics of formation of urea complexes with manganese (II), Nickel (II) and Zn (II) ions in N,N-dimethylformamide. Talanta 42:535-541

Palmer MR, Edmond JM (1993) Uranium in river water. Geochim Cosmochim Acta 57:4947-4955

Patterson MJ, Galloway SDR, Nimmo MA (2000) Variations in regional sweat composition in normal human males. Exp Physiol 85:869-875

Pietrzak-Flis Z, Rosiak L, Suplinska MM, Chrzanowski E, Dembinska S (2001) Daily intakes of ^{238}U, ^{234}U, ^{232}Th, ^{230}Th, ^{228}Th and ^{226}Ra in the adult population of the central Poland. Sci Total Environ 273:163-166

Prat O, Vercouter T, Ansoborlo E, Fichet P, Perret P, Kurttio P, Salonen L (2009) Uranium speciation in drinking water from drilled wells in southern Finland and its potential links to health effects. Environ Sci Technol 43:3941-3946

Prat O, Berenguer F, Steinmetz G, Ruat S, Sage N, Quemeneur E (2010) Alterations of gene expression in cultured human cells after acute exposure to uranium salt. Toxicol In Vitro 24: 160-168

Priest ND (2001) Toxicity of depleted uranium. Lancet 357:244-246

Puigdomenech I (2004) MEDUSA and HYDRA software for chemical equilibrium calculations. Royal Institute of Technology, Sweden

Raditzky B, Schmeide K, Sachs S, Geipel G, Bernhard G (2010) Interaction of uranium(VI) with nitrogen containing model ligands studied by laser-induced fluorescence spectroscopy. Polyhedron 29:620-626

Rajan KS, Martell AE (1965) Equilibrium studies of uranyl complexes. III. Interaction of uranyl ion with citric acid. Inorg Chem 4:462-469

Ramana VV, Thyagaraju VJ, Sastry KS (1992) Chromium Complexes of Uric Acid – Synthesis, Structure, and Properties. J Inorg Biochem 48:85-93

Robinson S, Robinson AH (1954) Chemical composition of sweat. Physiol Rev 34:202-220

Roth P, Werner E, Paretzke HG (2001) A study of uranium excreted in urine, An assessment of protective measures taken by the German Army KFOR Contingent. A Report for the Federal Ministry of Defense, GSF-National Research Centre for Environment & Health, Institute for Radiation Protection, Neuherberg, Germany

Runde W (2000) Spectroscopies for environmental studies of actinide species. Los Alamos Sci 26:412-415

Rutsch M, Geipel G, Brendler V, Bernhard G, Nitsche H (1999) Interaction of uranium(VI) with arsenate in aqueous solution studied by time-resolved laser-induced fluorescence sprectroscopy Radiochim Acta 86:135-141

Scapolan S, Ansoborlo E, Moulin C, Madic C (1997) Uranium Speciation in Biological Medium by Means of Capillary Electrophoresis and Time-Resolved Laser-Induced Fluorescence. J Radioanal Nucl Chem 226:145-148

Scapolan S, Ansoborlo E, Moulin C, Madic C (1998a) Investigations by time-resolved laser-induced fluorescence and capillary electrophoresis of the uranyl-phosphate species: application to blood serum. J Alloys Compd 271-273:106-111

Scapolan S, Ansoborlo E, Moulin C, Madic C (1998b) Uranium(VI)–transferrin system studied by time resolved laser-induced fluorescence. Radiat Prot Dosim 79:505-508

Schmeide K, Joseph C, Sachs S, Steudtner R, Raditzky B, Guenther A, Bernhard G (2012) Joint Project: Interaction and transport of actinides in natural clay rock with consideration of humic substances and clay organics – Characterization and quantification of the influence of clay organics on the interaction and diffusion of uranium and americium in the clay. Wissenschaflich-technische Berichte, HZDR-017, p25-30

Sharma A, Schulman SG (1999) Introduction to Fluorescence Spectroscopy. Wiley–Interscience, New York

Skoog DA, Leary JJ (1996) Instrumentelle Analytik. Springer, Berlin

Sparovek RBM, Fleckenstein J, Schnug E (2001) Issues of uranium and radioactivity in natural mineral waters. Landbauforschung Völkenrode 4:149-157

Song Z, Hou S (2002) Chemiluminescence assay for uric acid in human serum and urine using flow – injection with immobilized reagents technology. Anal Bioanal Chem 372:327-332

Sontag W (1986) Multicompartment kinetic models for the metabolism of americium, plutonium and uranium in rats. Hum Toxicol 5:163-173

Stern O, Volmer M (1919) Über die Abklingungszeiten der Fluoreszenz. Phys Z 20:183-188

Steudtner R, Arnold T, Großmann K, Geipel G, Brendler V (2006) Luminescence spectrum of Uranyl(V) in 2-propanol perchlorate solution. Inorg Chem Commun 9:939-941

Sutton M, Burastero SR (2003) Beryllium Chemical Speciation in Elemental Human Biological Fluids. Chem Res Toxicol 16:1145-1154

Sutton M, Burastero SR (2004) Uranium(VI) Solubility and Speciation in Simulated Elemental Human Biological Fluids. Chem Res Toxicol 17:1468-1480

Takeno N (2005) Atlas of Eh-pH Diagrams, Geological Survey of Japan, Open File Report No. 419.

Taylor DM, Taylor SK (1997) Environmental uranium and human health. Rev Environ Health 12:147-157

Templeton DM, Ariese F, Cornelis R, Danielsson L, Muntau H, van Leeuwen HP, Lobinski R (2000) Guidelines for terms related chemical speciation and fractionation of elements – Definitions, structural aspects, and methodological approaches. Pure Appl Chem 72:1453-1470

UBA (Umweltbundesamt) (2005) Uran und Human-Biomonitoring: Stellungnahme der Kommission Human-Biomonitoring des Umweltbundesamtes. Bundesgesundheitsblatt – Gesundheitsforschung – Gesundheitsschutz 48:822-827

UNSCEAR (United Nations Scientific Committee on the Effects of Atomic Radiation) (1993) Sources and effects of ionizing radiation. Report to the General Assembly of the United Nations with Scientific Annexes, United Nations sales publication E.94.IX.2, NNSCEAR, New York

UNSCEAR (United Nations Scientific Committee on the Effects of Atomic Radiation) (2000a) Sources and effects of ionizing radiation, Report to the general assembly with scientific annexes, vol. I. Sources, UNSCEAR, New York

UNSCEAR (United Nations Scientific Committee on the Effects of Atomic Radiation) (2000b) Sources and Effects of Ionising Radiation. Report to the General Assembly with Scientific Annexes. Vol. II. Effects, UNSCEAR, New York

US-EPA (United States – Environmental Protection Agency) (2000) 40 CFR parts 9, 141, and 142, National Primary Drinking Water Regulations; Radionuclides; Final Rule. Fed Regist 65:76708-76753

Valeur B (2001) Molecular Fluorescence: Principles and Applications. Wiley, New York

Van Staveren CJ, Fenton DE, Reinhoudt DN, van Eerden J, Harkema S (1987) Complexation of Urea and UO_2^{2+} in a Schiff Base Macrocycle: A Mimic of an Enzyme Binding Site. J Am Chem Soc 109:3456-3458

Venkatesan VK, Suryanarayana CV (1956) Conductance and Other Physical Properties of Urea Solutions. J Phys Chem 60:775-776

Vicente-Vicente L, Quiros Y, Perez-Barriocanal F, Lopez-Novoa JM, Lopez-Hernandez FJ, Morales AI (2010) Nephrotoxicity of uranium: pathophysiological, diagnostic and therapeutic perspectives. Toxicol Sci 118:324-347

Wang Z, Zachara JM, Yantasee W, Gassman PL, Liu C, Joly AG (2004) Cryogenic laser induced fluorescence characterization of U(VI) in Hanford Vadose zone pore waters. Environ Sci Technol 38:5591-5597

Weir E (2004) Uranium in drinking water, naturally. Can Med Assoc J 170:951-952

WHO (World Health Organization) (1998) Guidelines for Drinking-Water Quality – 2nd Edition – Vol. 2. Health Criteria and Other Supporting Information – Addendum (WHO/EOS/98.1). WHO, Geneva

WHO (World Health Organization) (2001) Depleted uranium: sources, exposure and health effects (WHO/SDE/PHE/01.1). Department of Protection of the Human Environment, WHO, Geneva

WHO (World Health Organization) (2003) Uranium in drinking-water. Background document for preparation of WHO Guidelines for drinking water quality (WHO/SDE/WSH/03.04/118). WHO, Geneva

WHO (World Health Organization) (2004) Uranium in drinking-water. Background document for development of WHO Guidelines for drinking water quality (WHO/SDE/WSH/03.04/118). WHO, Geneva

WHO (World Health Organization) (2011) Uranium in drinking-water. Background document for development of WHO Guidelines for drinking water quality (WHO/SDE/WSH/03.04/118/Rev/1). WHO, Geneva

Williams RJ, Lansford EM (1967) The Encyclopedia of Biochemistry. Reinhold Publishing Corp., New York, pp 66

Winde F (2011) Challenges in Assessing Uranium-Related Health Risks: Two Case Studies for the Aquatic Exposure Pathway from South Africa – Part I: Guideline and Toxicity Issues and the Pofadder Case Study. In: Merkel B, Schipek M (ed) The New Uranium Mining Boom. Springer-Verlag, Berlin Heidelberg, pp 529-538

Wolery TJ (1992) EQ3/6, A Software Package for the Geochemical Modeling of Aqueous Systems, UCRL-MA-110662 Part I. Lawrence Livermore National Laboratory, California

Wrenn ME, Durbin PW, Howard B, Lipsztein J, Rundo J, Sill ET, Willis DL (1985) Metabolism of ingested U and Ra. Health Phys 48:601-633

Yokoyama YU, Moriyasu M, Ikeda S (1976) Electron Transfer Mechanism in Quenching of Uranyl Luminescence by Halide Ions. J Inorg Nucl Chem 38:1329-1333

Yoshida H, Metcalfe R, Yamamoto K, Murakami Y, Hoshii D, Kanekiyo A, Naganuma T, Hayashi T (2008) Redox front formation in an uplifting sedimentary rock sequence: An analogue for redox-controlling processes in the geosphere around deep geological repositories for radioactive waste. Appl Geochem 23:2364-2381

Zamora ML, Tracy BL, Zielinski JM, Meyerhof DP, Moss MA (1998) Chronic ingestion of uranium in drinking water: a study of kidney bioeffects in humans. Toxicol Sci 43:68–77

Závodská L, Kosorínová E, Scerbáková L, Lesný J (2008) Environmental chemistry of uranium. HEJ Manuscript ENV-081221-A, HU ISSN 1418-7108, 1-18

Zheng ZP, Tokunaga TK, Wan JM (2003) Influence of calcium carbonate on U(VI) sorption to soils. Environ Sci Technol 37:5603-5608

Annexes

Annex I – Derivation of tolerable daily intake (TDI) and guideline value (GV) for uranium in drinking water

The TDI is defined as an estimate of the daily amount of a chemical over a lifetime without significant health risk. For uranium it is derived from the lowest value of which adverse health effects are observed in rats' kidneys at uptakes of 60 μg/Kg bw (body weight) per day as set by Gilman et al. (1998a).

$$TDI = \frac{60\ \mu g/Kg\ bw\, per\ day}{100} = 0.6\ \mu g/Kg\ bw\, per\ day$$

The 100 represents the uncertainty factor (× 10 for intraspecies variation and × 10 for interspecies variation).

According to the above calculated value, a health-based guideline value (GV) can be determined assuming a 70-Kg adult with average water consumption of 1.5 L/day.

$$GV = \frac{0.6\ \mu g/Kg\ bw\ per\ day \times 70\ Kg\ bw\ \times 0.35}{1.5\ L/day} = 10\ \mu g/L$$

The 0.35 in the calculations denotes the proportion of daily uranium intake allocated to drinking water.

This guideline value may change due to the variation in the daily water consumption as well as the allocation of the TDI to drinking water.

Annex II – Luminescence decay curves for U(VI) in all investigated mineral waters

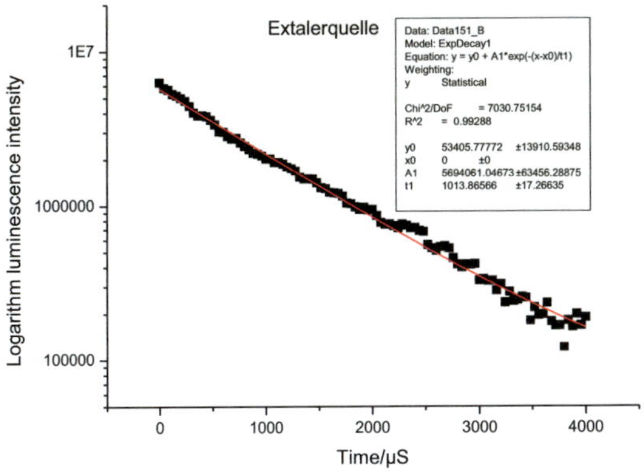

(a) Extalerquelle (STILL water)

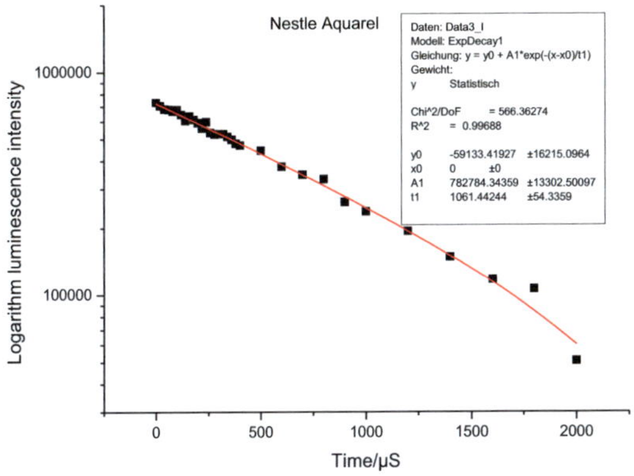

(b) Nestle Aquarel (STILL water)

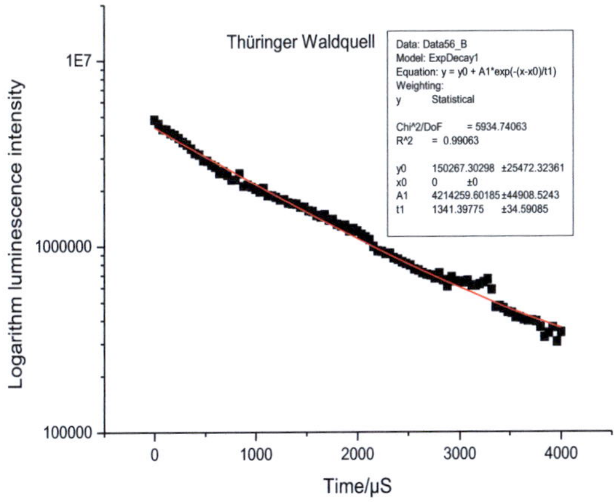

(c) Thüringer Waldquell (STILL water)

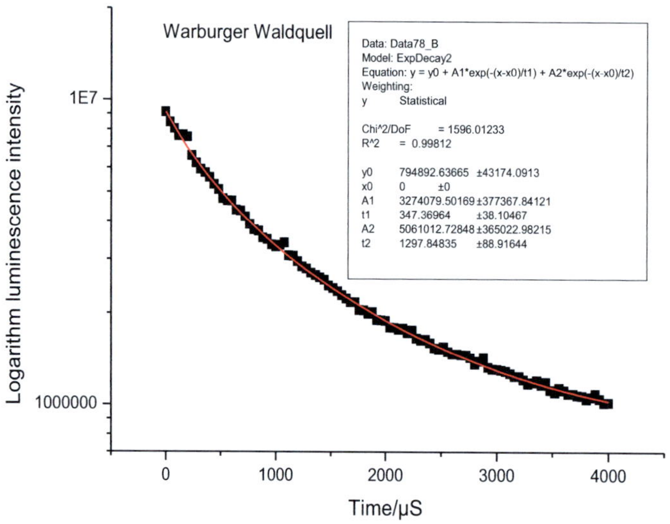

(d) Warburger Waldquell (CLASSIC water)

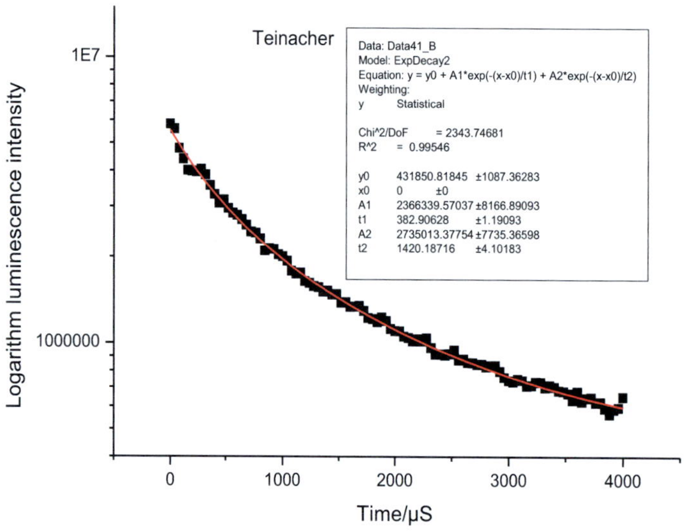

(e) Teinacher (CLASSIC water)

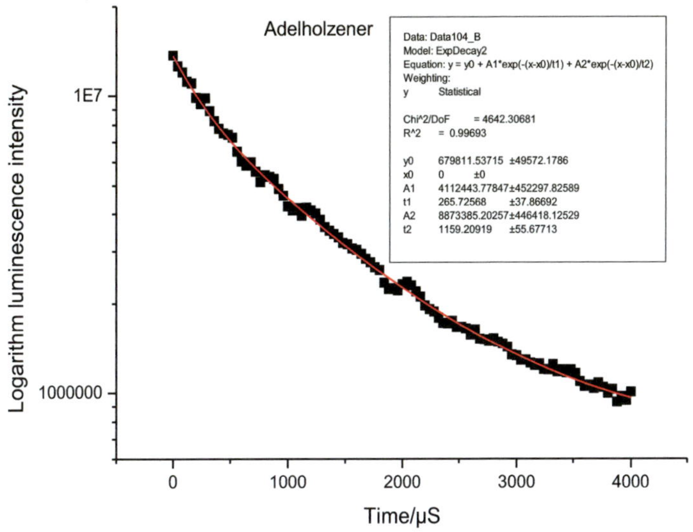

(f) Adelholzener (CLASSIC water)

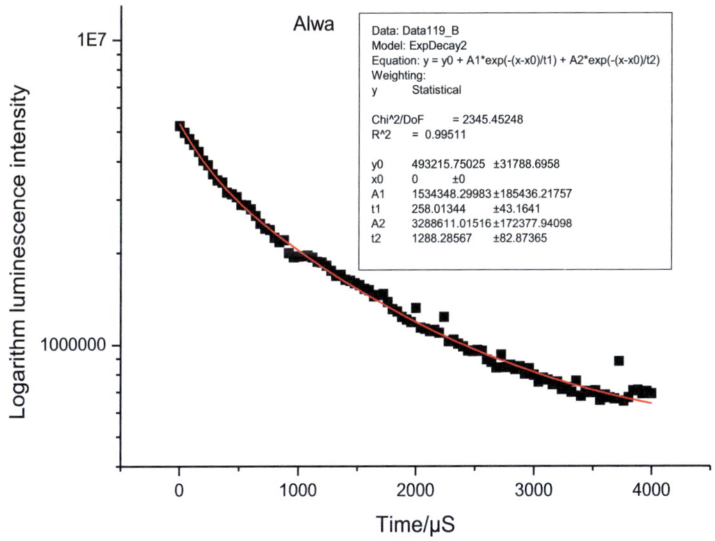

(g) Alwa (CLASSIC water)

Annex III – Luminescence decay curves for U(VI) in Pofadder groundwater

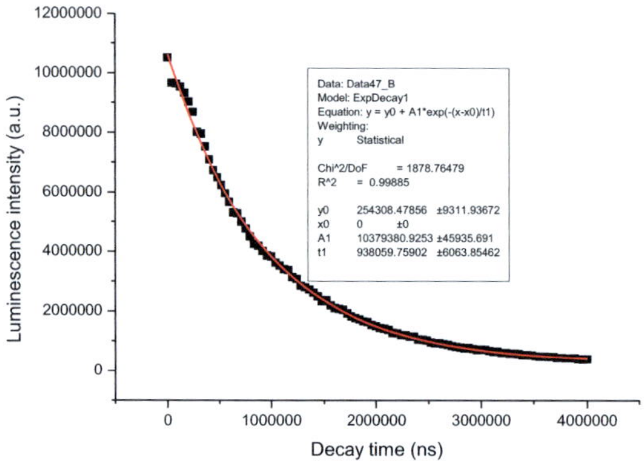

Spieelpan (S-pan)

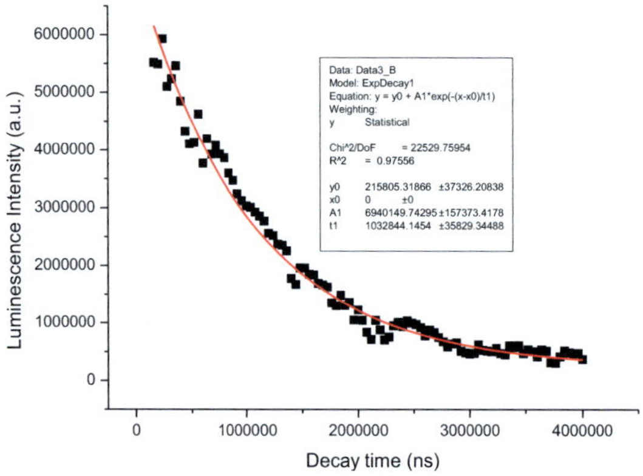

Poortijie (P)

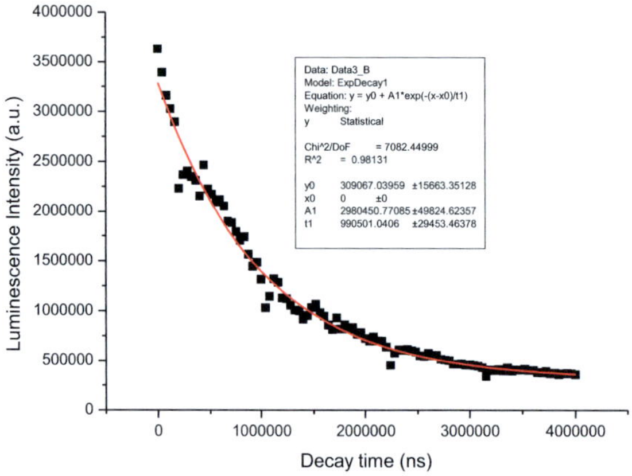

Potchefstroom (pot)

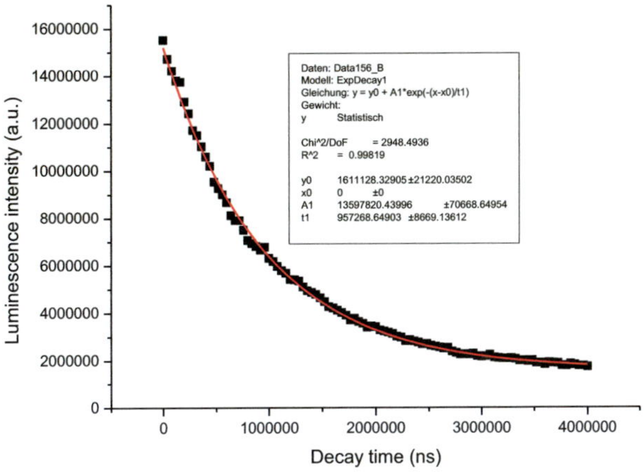

Calcrete wind pump (CC)

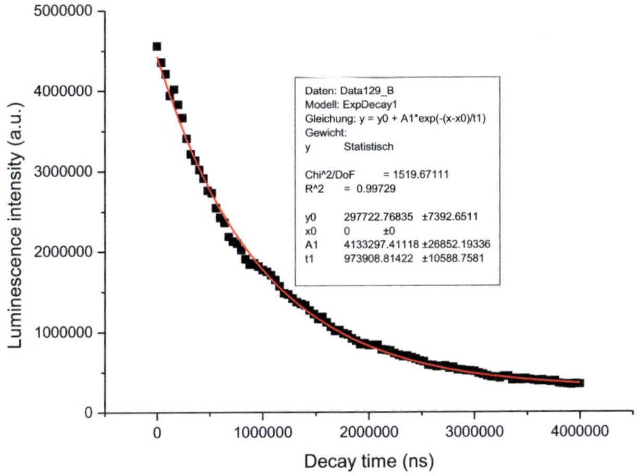

Grappies (GRP)

Annex IV – Luminescence decay curves for U(VI) in acid drainage waters

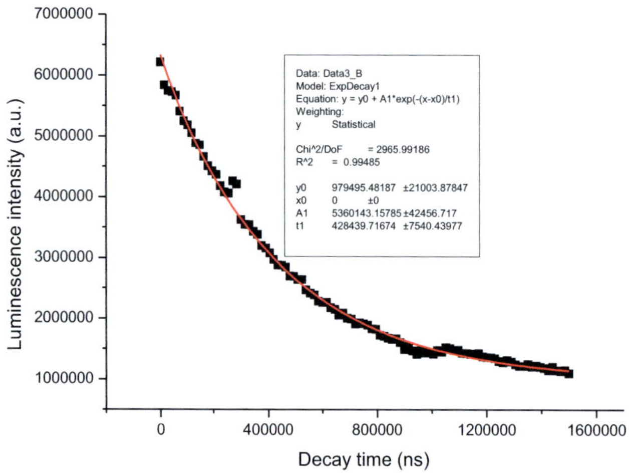

Wastewater – Kynoch

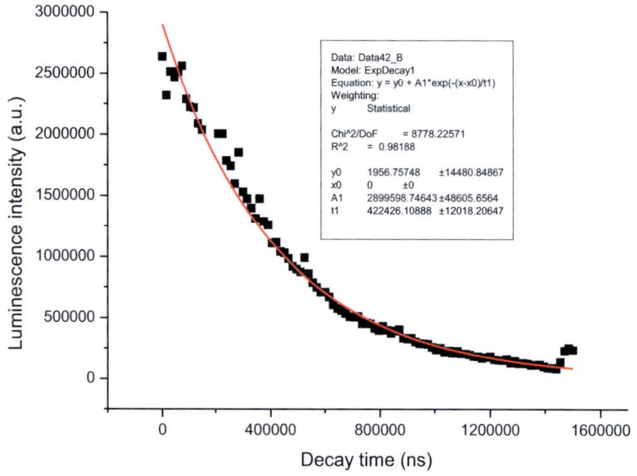

Mine water – Königstein

Annex V – Compositional analysis of the investigated body fluids as measured by elemental analysis techniques (ICP-MS and IC)

Table V(a). Aartificial body fluids

Element/ mM	**Artificial saliva**	**Artificial sweat**	**Artificial urine**
Na^+	14.7	67.8	132.2
K^+	11.6	6.36	70.3
Ca^{2+}	1.47	1	1.2
Mg^{2+}	0.19	0.4	2.5
Cl^-	12.7	69.7	130.9
SO_4^{2-}	0	0.4	18.3
PO_4^{3-}	6.4	0	12.3
CO_3^{2-}	3.11	9.4	5.95

Table V(b). Natural human saliva specimens

Element/ mM	**Specimen ID**			
	AO	CF	SF	Mean
Na^+	4.52	4.52	3.72	4.25
K^+	10.54	13.51	14.54	12.86
Ca^{2+}	1.18	1.29	1.23	1.23
Mg^{2+}	0.19	0.16	0.12	0.157
Cl^-	14.75	13.43	13.77	13.98
SO_4^{2-}	0.14	0.2	0.23	0.19
PO_4^{3-}	5.39	5.09	6.72	5.73
CO_3^{2-}	4.54	4.92	4.32	4.59
TOC^a	770	1000	1670	1146.7
pH^b	6.12	7.06	6.65	6.61
I^c	0.044	0.054	0.057	0.0516

^{a} in mg/L, ^{b} no unit, ^{c} in mol/L

Table V(c). Natural human sweat specimens

Element/ mM	Specimen ID			
	KM	CF	AO	Mean
Na^+	130.9	125.2	133.5	129.9
K^+	12.7	10.3	9.1	10.7
Ca^{2+}	1.32	1.18	0.69	1.06
Mg^{2+}	0.3	0.32	0.27	0.296
Cl^-	108.3	149.2	112.6	123.4
SO_4^{2-}	0.5	0.6	0.4	0.5
CO_3^{2-}	5.53	6.2	14.7	8.81
TOC^a	2390	1600	1400	1796.7
pH^b	6.9	7.3	7.53	7.24
I^c	0.132	0.159	0.160	0.15

Table V(d). Natural human urine specimens

Element/ mM	Specimen ID			
	SW	AO	AR	Mean
Na^+	103	132.6	109.1	114.9
K^+	156.4	91.5	79.7	109.2
Ca^{2+}	1.57	1.6	8.03	3.7
Mg^{2+}	1.97	3.6	6.6	4.1
Cl^-	177.7	159.1	135.1	157.3
SO_4^{2-}	7.45	15	20.6	14.4
PO_4^{3-}	16.1	22.6	22.4	20.4
CO_3^{2-}	7.9	0.4	0.5	2.9
TOC^a	5980	9480	12100	9186.7
pH^b	5.4	5.72	5.25	5.46
I^c	0.37	0.33	0.31	0.336

Annex VI – Luminescence decay curves and lifetimes for U(VI) spectra in natural human body fluids under investigation (Fig.VI(a, b, c))

Figure VI(a): Luminescence decay curves and lifetimes for U(VI) spectra in natural human saliva specimens (AO, CF, SF)

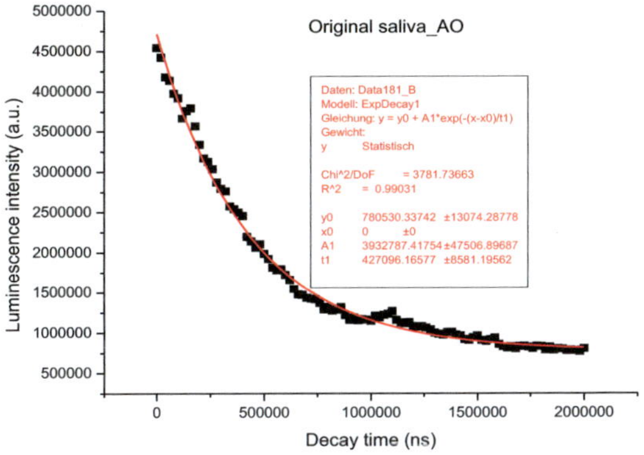

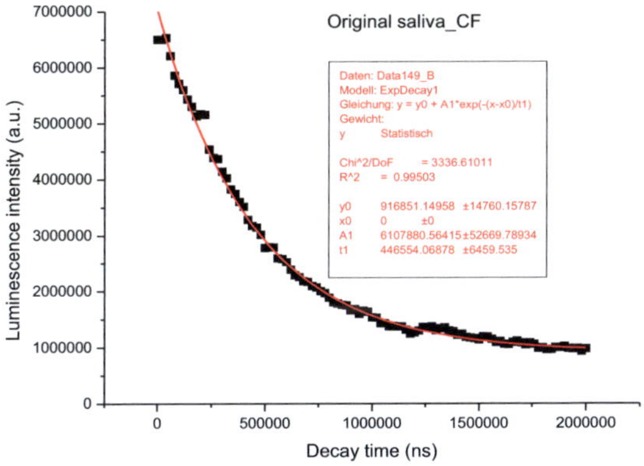

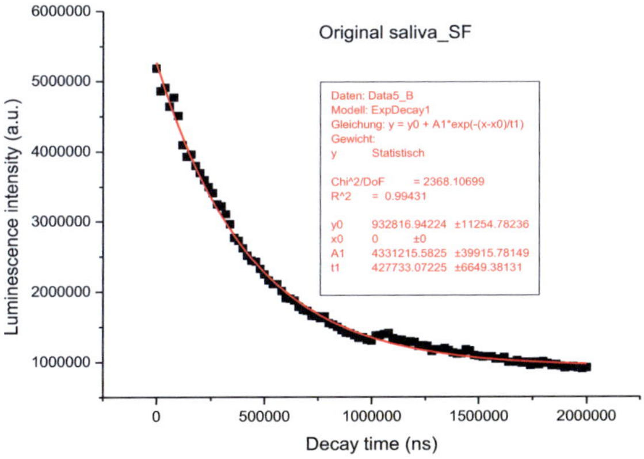

Figure VI(b): Luminescence decay curves and lifetimes for U(VI) spectra in natural human sweat specimens (KM, CF, AO)

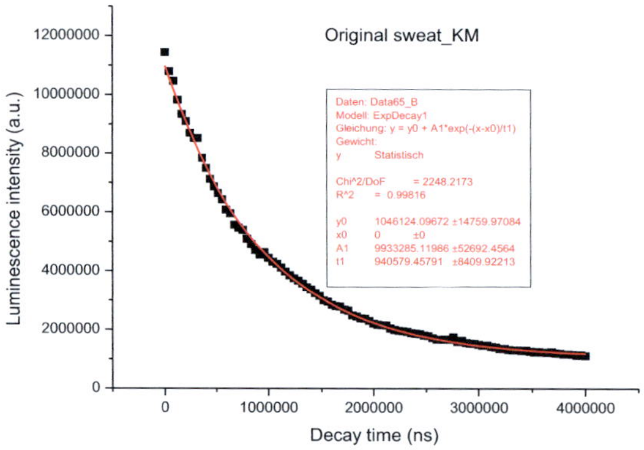

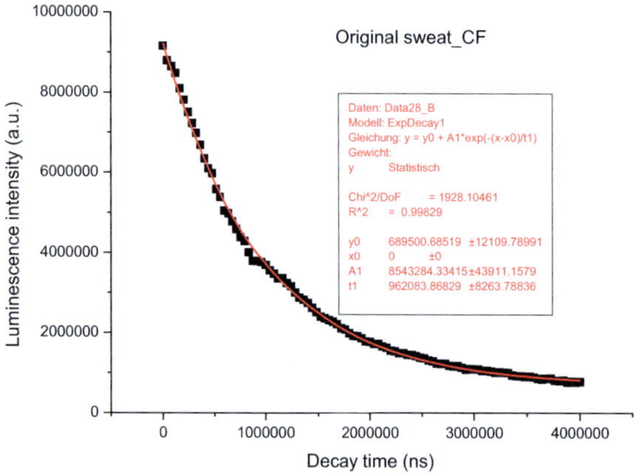

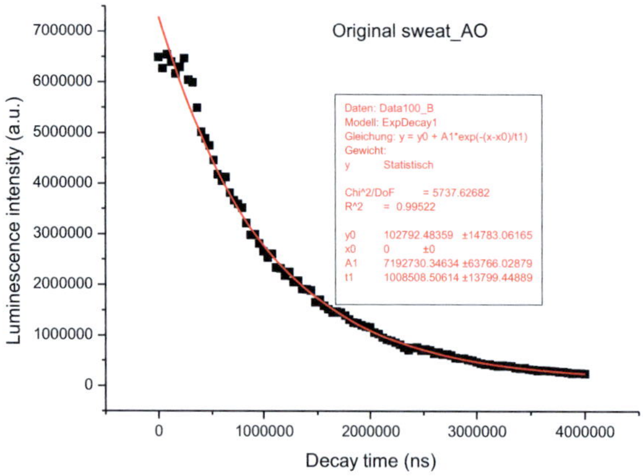

Figure VI(c): Luminescence decay curves and lifetimes for U(VI) spectra in natural human urine specimens (SW, AO, AR)

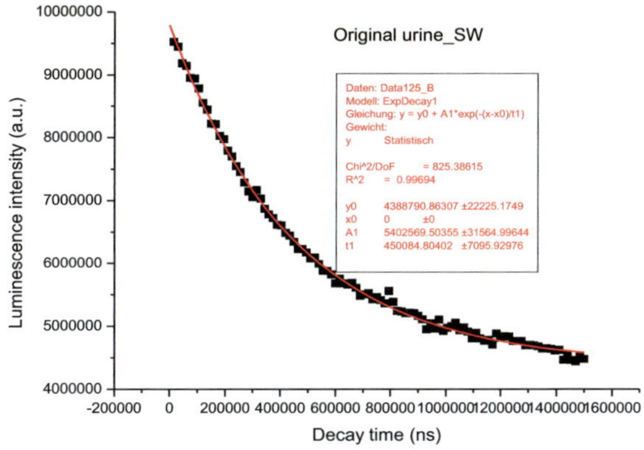

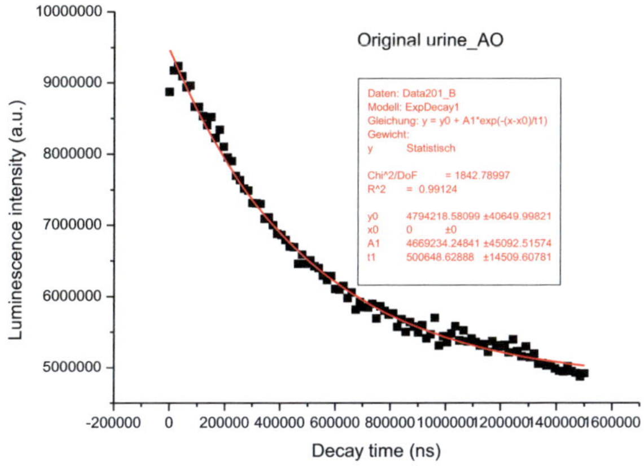

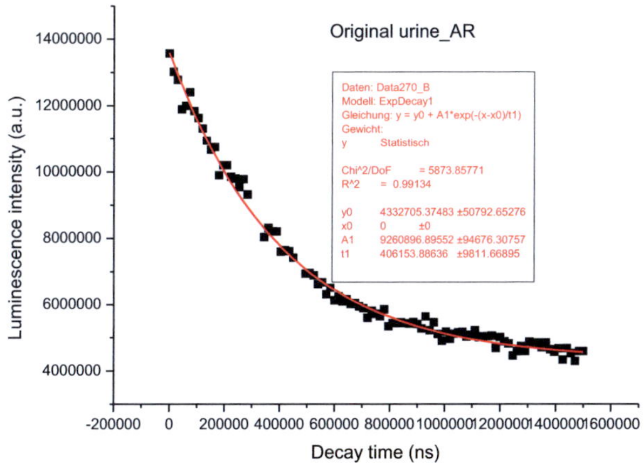

Annex VII – Photos: Equipment

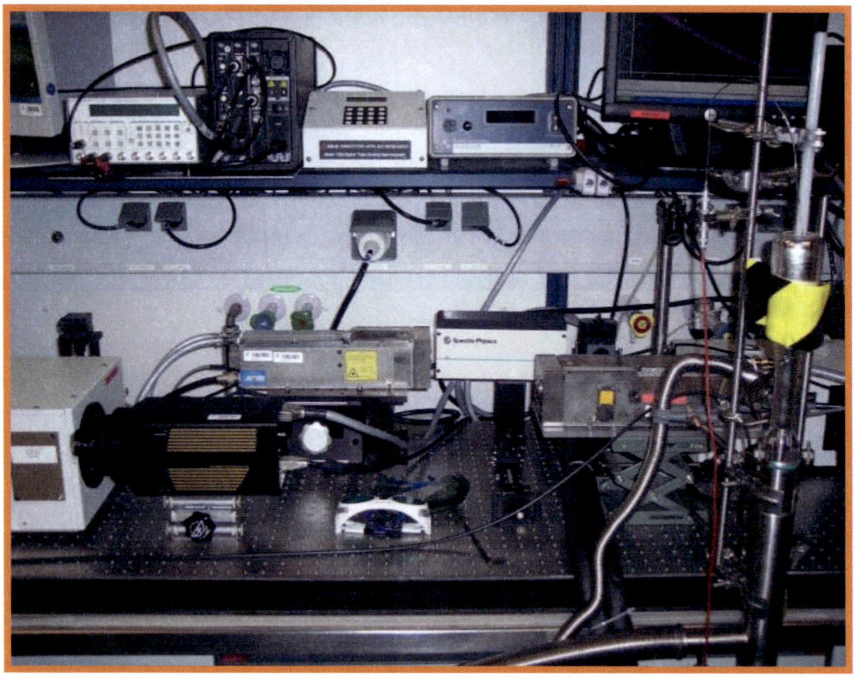

Laser fluorescence spectroscopy – setup

Cooling system and sample compartment

Eidesstattliche Erklärung

Hiermit versichere ich, dass ich die vorliegende Arbeit ohne unzulässige Hilfe Dritter und ohne Benutzung anderer als der angegebenen Hilfsmittel angefertigt habe; die aus fremden Quellen direkt oder indirekt übernommenen Gedanken sind als solche kenntlich gemacht. Die Arbeit wurde bisher weder im Inland noch im Ausland in gleicher oder ähnlicher Form einer anderen Prüfungsbehörde vorgelegt.

Die vorliegende Arbeit wurde am Helmholtz-Zentrum Dresden-Rossendorf e.V. am Institut für Ressourcenökologie in der Zeit von April 2010 bis Dezember 2013 unter der wissenschaftlichen Betreuung von Herrn Dr. G. Geipel und Herrn Prof. Dr. G. Bernhard angefertigt.

Ich erkenne die Promotionsordnung der Fakultät Mathematik und Naturwissenschaften der Technischen Universität Dresden in der Fassung vom 23.02.2011 an. Bisherige erfolglose Promotionsverfahren haben nicht stattgefunden.

Dresden, den 03.03.2014